AF710442

BASIC Programs for Steam Plant Engineers

MECHANICAL ENGINEERING

A Series of Textbooks and Reference Books

EDITORS

L. L. FAULKNER

Department of Mechanical Engineering
The Ohio State University
Columbus, Ohio

S. B. MENKES

Department of Mechanical Engineering
The City College of the
City University of New York
New York, New York

1. Spring Designer's Handbook, *by Harold Carlson*
2. Computer-Aided Graphics and Design, *by Daniel L. Ryan*
3. Lubrication Fundamentals, *by J. George Wills*
4. Solar Engineering for Domestic Buildings, *by William A. Himmelman*
5. Applied Engineering Mechanics: Statics and Dynamics, *by G. Boothroyd and C. Poli*
6. Centrifugal Pump Clinic, *by Igor J. Karassik*
7. Computer-Aided Kinetics for Machine Design, *by Daniel L. Ryan*
8. Plastics Products Design Handbook, Part A: Materials and Components; Part B: Processes and Design for Processes, *edited by Edward Miller*
9. Turbomachinery: Basic Theory and Applications, *by Earl Logan, Jr.*
10. Vibrations of Shells and Plates, *by Werner Soedel*
11. Flat and Corrugated Diaphragm Design Handbook, *by Mario Di Giovanni*
12. Practical Stress Analysis in Engineering Design, *by Alexander Blake*
13. An Introduction to the Design and Behavior of Bolted Joints, *by John H. Bickford*
14. Optimal Engineering Design: Principles and Applications, *by James N. Siddall*
15. Spring Manufacturing Handbook, *by Harold Carlson*
16. Industrial Noise Control: Fundamentals and Applications, *edited by Lewis H. Bell*
17. Gears and Their Vibration: A Basic Approach to Understanding Gear Noise, *by J. Derek Smith*

18. Chains for Power Transmission and Material Handling: Design and Applications Handbook, *by the American Chain Association*
19. Corrosion and Corrosion Protection Handbook, *edited by Philip A. Schweitzer*
20. Gear Drive Systems: Design and Application, *by Peter Lynwander*
21. Controlling In-Plant Airborne Contaminants: Systems Design and Calculations, *by John D. Constance*
22. CAD/CAM Systems Planning and Implementation, *by Charles S. Knox*
23. Probabilistic Engineering Design: Principles and Applications, *by James N. Siddall*
24. Traction Drives: Selection and Application, *by Frederick W. Heilich III and Eugene E. Shube*
25. Finite Element Methods: An Introduction, *by Ronald L. Huston and Chris E. Passerello*
26. Mechanical Fastening of Plastics: An Engineering Handbook, *by Brayton Lincoln, Kenneth J. Gomes, and James F. Braden*
27. Lubrication in Practice, Second Edition, *edited by W. S. Robertson*
28. Principles of Automated Drafting, *by Daniel L. Ryan*
29. Practical Seal Design, *edited by Leonard J. Martini*
30. Engineering Documentation for CAD/CAM Applications, *by Charles S. Knox*
31. Design Dimensioning with Computer Graphics Applications, *by Jerome C. Lange*
32. Mechanism Analysis: Simplified Graphical and Analytical Techniques, *by Lyndon O. Barton*
33. CAD/CAM Systems: Justification, Implementation, Productivity Measurement, *by Edward J. Preston, George W. Crawford, and Mark E. Coticchia*
34. Steam Plant Calculations Manual, *by V. Ganapathy*
35. Design Assurance for Engineers and Managers, *by John A. Burgess*
36. Heat Transfer Fluids and Systems for Process and Energy Applications, *by Jasbir Singh*
37. Potential Flows: Computer Graphic Solutions, *by Robert H. Kirchhoff*
38. Computer-Aided Graphics and Design, Second Edition, *by Daniel L. Ryan*
39. Electronically Controlled Proportional Valves: Selection and Application, *by Michael J. Tonyan, edited by Tobi Goldoftas*
40. Pressure Gauge Handbook, *by AMETEK, U.S. Gauge Division, edited by Philip W. Harland*
41. Fabric Filtration for Combustion Sources: Fundamentals and Basic Technology, *by R. P. Donovan*
42. Design of Mechanical Joints, *by Alexander Blake*
43. CAD/CAM Dictionary, *by Edward J. Preston, George W. Crawford, and Mark E. Coticchia*

44. Machinery Adhesives for Locking, Retaining, and Sealing, *by Girard S. Haviland*
45. Couplings and Joints: Design, Selection, and Application, *by Jon R. Mancuso*
46. Shaft Alignment Handbook, *by John Piotrowski*
47. BASIC Programs for Steam Plant Engineers: Boilers, Combustion, Fluid Flow, and Heat Transfer, *by V. Ganapathy*
48. Solving Mechanical Design Problems with Computer Graphics, *by Jerome C. Lange*
49. Plastics Gearing: Selection and Application, *by Clifford E. Adams*
50. Clutches and Brakes: Design and Selection, *by William C. Orthwein*
51. Transducers in Mechanical and Electronic Design, *by Harry L. Trietley*

ADDITIONAL VOLUMES IN PREPARATION

Mechanical Engineering Software

1. Spring Design with an IBM PC, *by Al Dietrich*

BASIC Programs for Steam Plant Engineers

Boilers, Combustion, Fluid Flow, and Heat Transfer

V. Ganapathy

ABCO Industries, Inc.
Abilene, Texas

MARCEL DEKKER, INC. New York and Basel

Library of Congress Cataloging-in-Publication Data

Ganapathy, V.
BASIC programs for steam plant engineers.

(Mechanical engineering ; 47)
Includes index.
1. Steam power-plants--Computer programs. 2. BASIC (Computer program language) 3. Microcomputers--Programming. I. Title. II. Series.
TJ403.G36 1986 621.1'8'0285526 86-2110
ISBN 0-8247-7489-2

COPYRIGHT © 1986 by MARCEL DEKKER, INC. ALL RIGHTS RESERVED

Neither this book nor any part may be reproduced or transmitted in any form or by any means, electronic or mechanical, including photocopying, microfilming, and recording, or by any information storage and retrieval system, without permission in writing from the publisher.

MARCEL DEKKER, INC.
270 Madison Avenue, New York, New York 10016

Current printing (last digit):
10 9 8 7 6 5 4 3 2 1

PRINTED IN THE UNITED STATES OF AMERICA

This book is dedicated to the memory of my father,

P. Viswanatha Iyer

Preface

Over the past ten years, I have had the opportunity to author over one hundred short articles and three books of interest to steam plant engineers, covering such topics as heat transfer, heat recovery equipment design, boiler calculations, fluid flow, and steam plant equipment sizing. With the plethora of hand-held and personal computers, I thought that it would be helpful to steam plant engineers if several of the shorter and more frequently made calculations could be performed by computer. The result is this volume, which has thirty useful programs written in BASIC. The programs run directly on the IBM PC and compatible systems. The programs can also be modified to run on hand-held computers such as the Sharp, TRS-80, and Texas Instruments models.

These programs are based on my years of experience, having engineered the systems discussed in the text. In contrast, many programs on the market are developed by software engineers who are experts on programming, but lack sufficient knowledge of steam plants and processes and the needs of design and operating engineers. For instance, the programs on combustion calculations and boiler efficiency take into account the effect of relative humidity, an important factor which could easily be overlooked by engineers with inadequate engineering experience. Flue gas analysis is reported on a dry and wet basis, which is helpful in figuring excess air, flue gas density, and gas properties. Efficiency of boilers is determined using ASME PTC procedures and is reported on a lower as well as on a higher heating value basis.

The programs on calculation of pressure drop and heat transfer coefficients inside tubes require only easily available data such as flow per tube, fluid pressure, temperature, and pipe size. Reference to steam tables or properties of air, water, and flue gas is avoided, enabling field engineers to obtain the data with ease, thus saving considerable time. Predicted pressure drop may be compared with the measured value to see if the equipment malfunctioned.

In recognition of the widespread use of finned tubes in boilers and heaters, a number of programs are presented to predict heat transfer and pressure drop in these types of equipment. Various fin geometries can be evaluated to study alternates.

Furnaces and heat transfer equipment are routinely lined with several layers of insulation. Engineers are often required to predict temperature profiles across the various layers and heat loss under varying conditions of ambient temperature and wind velocity. Program 3.9 performs this calculation, handling any number of layers.

Steam properties after throttling or after expansion are important to those involved in valve and pipe sizing and steam turbine selection. Programs 4.1 and 4.2 perform these involved calculations. Steam blowing is a routinely performed operation in boiler plants. Program 4.5 predicts the steam flow rate during a sonic flow situation, when steam escapes to atmosphere.

Estimating the performance of fire tube and water tube waste heat boilers and economizers is vital for engineers involved in their design and operation. Programs 5.1 through 5.3 predict the exit gas temperature and duty at any inlet conditions of gas flow and temperature. Plant engineers can then check to see whether their equipment is functioning properly.

The data required by these programs are easily available to all practicing design and field engineers. The results are practical and useful information for design and performance evaluation of steam plant equipment. Each program listing includes theory, correlations used, logic, listing, and examples, along with a printout of results. In addition, a table of nomenclature has been included with each program. This helps relate the variables used in the text with those in the computer so the programs can be modified to run on other computers. One can also obtain the value of any variable by pressing the appropriate key. A look at the brief synopsis at the beginning of each program will reveal to the user the amount of practical information that can be obtained.

The aim of this volume is to present workable BASIC programs of interest to steam plant designers, consultants, and operators. As with any project, there is always room for some improvement. Therefore, comments from readers on the style and approach of the programs are always welcome.

V. Ganapathy

Diskette to accompany *BASIC Programs for Steam Plant Engineers*: I have prepared a diskette (for IBM PC and compatibles) containing all of the programs discussed in this book. If you would like to purchase a copy, or need more information, please contact me at the following address:
V. Ganapathy, P.O. Box 673, Abilene, TX 79604

Contents

1

Fuels, Combustion, and Efficiency of Boilers and Heaters

PROGRAM 1.1 COMBUSTION CALCULATIONS FOR SOLID AND LIQUID FUELS

Input

Carbon
Hydrogen
Oxygen
Nitrogen
Sulfur
Moisture
Excess air
Ambient temperature
Relative humidity

Output

Wet flue gas analysis
Dry flue gas analysis
Molecular weight of flue gas
Density of flue gas
Dry air required per pound of fuel
Wet air per pound of fuel
Dry flue gas produced per pound of fuel
Wet flue gas produced per pound of fuel
Higher and lower heating values of the fuel

Remarks

Results are helpful in predicting the performance of fired heat transfer equipment.

Given the ultimate analysis of solid or liquid fuels, excess air, ambient temperature, and relative humidity, this program performs detailed combustion calculations and prints out the dry and wet flue gas analysis, dry and wet air quantities, and dry and wet flue gas produced, along with the flue gas molecular weight and density. These data are useful for heat transfer calculations and to estimate performance of boilers, heaters, and combustion related equipment. Heating values of the fuel are also printed out.

From partial pressure of carbon dioxide and water vapor, one can estimate nonluminous heat transfer coefficients. Dry fuel gas analysis helps the engineer to figure excess air for combustion

TABLE 1.1A Equations Used for Combustion Calculations

$$SVP = 0.08 + 281 \times 10^{-9} T^{3.25} \qquad (1)$$

$$M = \frac{0.622 \times SVP \times RH}{14.7 - SVP \times RH} \qquad (2)$$

$$w_{da} = 100(2.664C + 7.937H_2 + S - O_2)\frac{E}{23} \qquad (3)$$

$$w_{wa} = w_{da}(1 + M) \qquad (4)$$

$$w_{wg} = \frac{3.66C + 8.94H_2 + W + w_{da}M + 2S + 0.77w_{da} + N_2 + 0.23w_{da}(E-1)}{E} \qquad (5)$$

$$w_{dg} = w_{wg} - (8.94H_2 + W + w_{da}M) \qquad (6)$$

$$F = 0.08318C + 0.03125S + \frac{8.94H_2 + W + w_{da}M)}{18} + \frac{0.77w_{da} + N_2}{28} + \frac{0.23w_{da}(E-1)}{32E} \qquad (7)$$

$$CO_2' = \frac{8.318C}{F} \qquad (8)$$

$$O_2' = \frac{23w_{da}(E-1)}{32EF} \qquad (9)$$

$$N_2' = \frac{0.77w_{da} + N_2}{28F} 100 \qquad (10)$$

$$H_2O' = \frac{8.94H_2 + W + w_{da}M}{18} 100 \qquad (11)$$

$$SO_2 = \frac{3.125S}{F} \qquad (12)$$

$$MW = \frac{44CO_2' + 32O_2' + 28N_2' + 18H_2O' + 64SO_2'}{100} \qquad (13)$$

$$\rho_g = 0.002636MW \qquad (14)$$

TABLE 1.1A (Continued)

$$O'_{2\,d} = \frac{O'_2}{100 - H_2O'}\,100 \tag{15}$$

$$CO'_{2\,d} = \frac{CO'_2}{100 - H_2O'}\,100 \tag{16}$$

$$N'_{2\,d} = \frac{N'_2}{100 - H_2O'}\,100 \tag{17}$$

$$\mathrm{HHV} = 14{,}500C + 62{,}000\left(H_2 - \frac{O_2}{8}\right) + 4000S \tag{18}$$

$$\mathrm{LHV} = \mathrm{HHV} - 9720H_2 - 1110W \tag{19}$$

control purposes. Knowing the density of flue gas aids in estimation of flue gas velocity in ducts.

Table 1.1A shows the equations used for performing the combustion calculations. Table 1.1B gives the nomenclature used, and Fig. 1.1 shows the listing of the program along with the results. An example illustrates the use of the program.

Example

A coal fired boiler operates under the following fuel conditions (fraction by weight):

Carbon = 0.728
Hydrogen = 0.048
Oxygen = 0.062
Nitrogen = 0.015
Sulfur = 0.022
Moisture = 0.035, rest ash
Excess air = 25%
Ambient temperature = 80°F
Relative humidity = 0.60 (60%)

Perform detailed combustion calculations.

TABLE 1.1B Nomenclature Used in Combustion Calculations

Nomenclature	Program symbol	Description and units
CO_2'	I	%Carbon dioxide in wet flue gas
O_2'	J	%Oxygen in wet flue gas
N_2'	K	%Nitrogen in wet flue gas
H_2O'	L	%Water vapor in wet flue gas
MW	A35	Molecular weight of flue gas
SO_2'	Q	%Sulfur dioxide in wet flue gas
M	M	Moisture in air (lb/lb fuel)
SVP	P	Saturated vapor pressure (psia)
T	T	Ambient temperature (°F)
RH	R	Relative humidity (fraction)
ρ_g	D	Gas denisty at 60°F (lb/ft^3)
O_{2d}	U	%Oxygen in dry flue gas
CO_{2d}	V	%Carbon dioxide in dry flue gas
N_{2d}	A45	%Nitrogen in dry flue gas
C	C	Carbon in fuel (fraction)
H_2	H	Hydrogen in fuel (fraction)
S	S	Sulfur in fuel (fraction)
O_2	O	Oxygen in fuel (fraction)
N_2	N	Nitrogen in fuel (fraction)
F	F	Moles of wet flue gas
W	W	Moisture in fuel (fraction)
w_{da}	A	Dry air for combustion (lb/lb fuel)
w_{wa}	B	Wet air for combustion (lb/lb fuel)
w_{dg}	Z	Dry flue gas (lb/lb fuel)
w_{wg}	G	Wet flue gas (lb/lb fuel)
E	E	Excess air factor; 1.25 means 25%
HHV	HHV	Higher heating value (Btu/lb)
LHV	LHV	Lower heating value (Btu/lb)

```
10  PRINT"COMBUSTION CALCULATIONS-SOLID,LIQUID FUELS"
20 INPUT"CARBON,HYDROGEN,OXYGEN,NITROGEN,SULFUR,MOISTURE,(fractions),EXCESS AIR%
=";C,H,O,N,S,W,E
25 INPUT"RELATIVE HUMIDITY(fraction),AMBIENT TEMP=";R,T
30 P=.08+281*10^-9*T^3.25:M=.622*P*R/(14.7-P*R)
35 HHV=14500*C+62000!*(H-.125*O)+4000*S:LHV=HHV-9720*H-1110*W
40 A=(2.664*C+7.937*H+S-O)*(100+E)/23:B=A*(1+M)
50 A30=8.939999*H+W+A*M:G=3.66*C+A30+2*S+.77*A+N+.23*A*E/(100+E):Z=G-A30
60 F=.08318*C+.03125*S+(A30/18)+((.77*A+N)/28)+.23*A*E/32/(100+E)
70 I=8.318*C/F:J=23*A*E/32/F/(100+E):K=((.77*A+N)/(28*F))*100:L=(A30/18/F)*100
80 Q= (.03125*S/F)*100:A35=(44*I+32*J+28*K+18*L+64*Q)/100:D=.002636*A35
90 U=100*(J/(100-L)):V=100*(I/(100-L)):A45=100*(K/(100-L))
95 PRINT" "
100 PRINT"RESULTS-COMBUSTION CALCULATIONS"
105 PRINT" "
110 PRINT"FUEL DATA-fractions"
120 PRINT"CARBON,HYDROGEN,OXYGEN,NITROGEN,SULFUR,MOISTURE"
130 PRINT USING"###.###";C,H,O,N,S,W
140 PRINT "REL-HUM-fraction=";R;"EXCESS AIR % =";E
145 PRINT" "
150 PRINT"WET FLUE GAS ANALYSIS %"
160 PRINT"CARBON DIOXIDE=";I;"OXYGEN=";J;"NITROGEN=";K
170 PRINT"SULFUR DIOXIDE=";Q;"WATER VAPOR=";L
180 PRINT" "
190 PRINT"AIR-GAS QUANTITIES-Lb/Lb fuel"
200 PRINT"DRY AIR REQD=";A;"WET AIR REQD=";B
210 PRINT"DRY GAS PRODUCED=";Z;"WET GAS PRODUCED=";G
220 PRINT" "
230 PRINT"DRY FLUE GAS ANALYSIS  %"
240 PRINT"CARBON DIOXIDE=";V;"OXYGEN=";U;"NITROGEN=";A45
250 PRINT" "
260 PRINT"MOLECULAR WEIGHT=";A35;"FLUE GAS DENSITY,Lb/cu ft-60 F=";D
265 PRINT" "
270 PRINT"HEATING VALUES-BTU/LB:HHV=";HHV;"LHV=";LHV
275 PRINT" "
280 END
```

```
RESULTS-COMBUSTION CALCULATIONS

FUEL DATA-fractions
CARBON,HYDROGEN,OXYGEN,NITROGEN,SULFUR,MOISTURE
  0.728  0.048  0.062  0.015  0.022  0.035
REL-HUM-fraction= .6 EXCESS AIR % = 25

WET FLUE GAS ANALYSIS %
CARBON DIOXIDE= 13.29994 OXYGEN= 3.912859 NITROGEN= 74.97235
SULFUR DIOXIDE= .1509983 WATER VAPOR= 7.663852

AIR-GAS QUANTITIES-Lb/Lb fuel
DRY AIR REQD= 12.39331 WET AIR REQD= 12.55727
DRY GAS PRODUCED= 12.83642 WET GAS PRODUCED= 13.46451

DRY FLUE GAS ANALYSIS  %
CARBON DIOXIDE= 14.40382 OXYGEN= 4.237624 NITROGEN= 81.19501

MOLECULAR WEIGHT= 29.57248 FLUE GAS DENSITY,Lb/cu ft-60 F= 7.795305E-02

HEATING VALUES-BTU/LB:HHV= 13139.5 LHV= 12634.09
```

FIG. 1.1 Listing of program 1.1, with results.

Solution

Key in the program, a listing of which is shown in Fig. 1.1. In the RUN mode, the screen asks for the data in the same order; they are fed in. Then the computer goes on to solve the equations shown in Table 1.1A and prints out the results.

It is seen that the wet flue gas produced is 13.464 lb/lb fuel, wet air required is 12.557 lb/lb fuel, and flue gas density is 0.077 lb/ft^3 at 60°F. Dry and wet gas analysis are also obtained.

Any data or result may be obtained by pressing the appropriate key (see Table 1.1B). Heating values of the fuel are also printed out. It is seen that the higher heating value is 13,139 Btu/lb.

Reference

V. Ganapathy, *Applied Heat Transfer*, Pennwell Books, Tulsa, Oklahoma, 1982, pp. 5-20.

PROGRAM 1.2 COMBUSTION CALCULATIONS FOR GASEOUS FUELS

Input

Volume of constituents of gaseous fuel
Excess air
Ambient temperature
Relative humidity

Output

Higher heating value (Btu/ft^3)
Higher heating value (Btu/lb)
Lower heating value (Btu/ft^3)
Lower heating value (Btu/lb)
Molecular weight of fuel
Molecular weight of flue gas
Dry and wet flue gas analysis
Wet air required for combustion
Wet flue gas produced
Water dew point

Remarks

Up to 19 gases can be handled.

This program performs detailed combustion calculations for gaseous fuels and obtains such data as

- Molecular weights of fuel and flue gas
- Wet and dry flue gas analysis
- Wet and dry air for combustion
- Water dew point
- Wet flue gas quantity
- Moles of air and flue gas per mole of fuel
- Heating values (higher and lower) of fuel (Btu/lb and Btu/ft^3)

The program considers the effect of moisture in air and relative humidity, which results in accurate flue gas analysis required for boiler efficiency determination. In all, 19 gases can be handled (see Table 1.2A).

The equations used are as follows: the basis is 1 mol of fuel.

Theory

Theoretical dry air (ft^3/ft^3 of fuel) v_t is given by

$$v_t = \sum \frac{a_i y_i}{100} \tag{1}$$

The molecular weight of fuel is

$$MW_f = \sum \frac{MW_i y_i}{100} \tag{2}$$

The actual dry air v_a, based on excess air of E%, is

$$v_a = v_t \left[1 + \left(\frac{E}{100}\right)\right] \tag{3}$$

The amount (mols) of CO_2 produced is

$$v_{CO_2} = \sum b_i y_i/100 \tag{4}$$

TABLE 1.2A Fuel Key[a]

Fuel number	Program symbol	Description
1	A(1)	CH_4, methane
2	A(2)	C_2H_6, ethane
3	A(3)	C_3H_8, propane
4	A(4)	C_4H_{10}, butane
5	A(5)	C_5H_{12}, pentane
6	A(6)	C_2H_4, ethylene
7	A(7)	C_3H_6, propylene
8	A(8)	C_4H_8, butylene
9	A(9)	C_6H_6, benezene
10	A(10)	C_7H_8, toluene
11	A(11)	C_2H_2, acetylene
12	A(12)	NH_3, ammonia
13	A(13)	H_2S, hydrogen sulfide
14	A(14)	H_2O, water vapor
15	A(15)	N_2, nitrogen
16	A(16)	CO_2, carbon dioxide
17	A(17)	CO, carbon monoxide
18	A(18)	H_2, hydrogen
19	A(19)	SO_2, sulfur dioxide

[a]All constituents are in % volume.

The (mols) of H_2O produced is

$$v_w = \sum c_i y_i/100 + v_m v_a \tag{5}$$

where v_m = 100 moles of water vapor in air per mole of dry air and

$$v_m = \frac{SVP \times RH}{(14.7 - SVP \times RH)} \tag{6}$$

where SVP is saturation vapor pressure

$$SVP = 0.08 + 0.281 \times 10^{-6} t^{3.25} \tag{7}$$

where t is the ambient temperature (°F) and RH is the relative humidity (fraction); y_i is the volume % of each constituent. a_i, b_i, and c_i are constants depending on fuel type. For methane, for example, a - 9.528, b - 1, and c - 2. (See Tables 1.2B and 1.2C.)

Flue gas analysis is obtained by solving the following equations. The amount (mol) of N_2 in flue gas:

$$v_{N2} = 0.79v_a + 0.01 \times \%N_2 \text{ in fuel} \tag{8}$$

The amount (mol) of SO_2:

$$v_{SO2} = 0.01(\%SO_2 + \%H_2S \text{ in fuel}) \tag{9}$$

The amount (mol) of oxygen in flue gas:

$$v_{O2} = \frac{0.0021v_a}{1 + E/100} \tag{10}$$

Once the various products are obtained, the total volume of flue gas per mole of fuel is obtained as follows:

$$v_g = v_{SO_2} + v_{O_2} + v_{N_2} + v_w + v_{CO_2} \tag{11}$$

Wet flue gas analysis may be obtained as follows:

$$p_c = \%CO_2 \text{ in flue gas} = 100\ (v_{CO_2}/v_g) \tag{12}$$

$$p_o = \%O_2 \text{ in flue gas} = 100\ (v_{O_2}/v_g) \tag{13}$$

$$p_w = \%H_2O \text{ in flue gas} = 100\ (v_w/v_g) \tag{14}$$

$$p_n = \%N_2 \text{ in flue gas} = 100\ (v_{N2}/v_g) \tag{15}$$

$$p_s = \%SO_2 \text{ in flue gas} = 100\ (v_{SO2}/v_g) \tag{16}$$

Dry flue gas analysis is obtained by multiplying each constituent by the factor $[100/(100 - p_w)]$.

The water dew point t_d is obtained from

$$t_d = [(0.147p_w - 0.08) \times 10^6/0.281]^{0.3077} \tag{17}$$

TABLE 1.2B Combustion Constants

No.	Substance	Formula	Molecular Weight[a]	Lb per Cu Ft[b]	Cu Ft per Lb[b]	Sp Gr Air = 1.000[b]	Heat of Combustion[c]: Btu per Cu Ft Gross	Btu per Cu Ft Net[d]	Btu per Lb Gross	Btu per Lb Net[d]	Cu Ft per Cu Ft of Combustible: Required for Combustion O_2	Required N_2	Required Air	Flue Products CO_2	Flue Products H_2O	Flue Products N_2	Lb per Lb of Combustible: Required for Combustion O_2	Required N_2	Required Air	Flue Products CO_2	Flue Products H_2O	Flue Products N_2	Experimental Error in Heat of Combustion Percent + or −
1	Carbon	C	12.01	—	—	—	—	—	14,093[g]	14,093[g]	—	—	—	—	—	—	2.664	8.863	11.527	3.664	—	8.863	0.012
2	Hydrogen	H_2	2.016	0.005327	187.723	0.06959	325.0	275.0	61,100	51,623	0.5	1.882	2.382	—	1.0	1.882	7.937	26.407	34.344	—	8.937	26.407	0.015
3	Oxygen	O_2	32.000	0.08461	11.819	1.1053	—	—	—	—	—	—	—	—	—	—	—	—	—	—	—	—	—
4	Nitrogen (atm)	N_2	28.016	0.07439[e]	13.443[e]	0.9718[e]	—	—	—	—	—	—	—	—	—	—	—	—	—	—	—	—	—
5	Carbon monoxide	CO	28.01	0.07404	13.506	0.9672	321.8	321.8	4,347	4,347	0.5	1.882	2.382	1.0	—	1.882	0.571	1.900	2.471	1.571	—	1.900	0.045
6	Carbon dioxide	CO_2	44.01	0.1170	8.548	1.5282	—	—	—	—	—	—	—	—	—	—	—	—	—	—	—	—	—
	Paraffin series C_nH_{2n+2}																						
7	Methane	CH_4	16.041	0.04243	23.565	0.5543	1013.2	913.1	23,879	21,520	2.0	7.528	9.528	1.0	2.0	7.528	3.990	13.275	17.265	2.744	2.246	13.275	0.033
8	Ethane	C_2H_6	30.067	0.08029[e]	12.455[e]	1.04882[e]	1792	1641	22,320	20,432	3.5	13.175	16.675	2.0	3.0	13.175	3.725	12.394	16.119	2.927	1.798	12.394	0.030
9	Propane	C_3H_8	44.092	0.1196[e]	8.365[e]	1.5617[e]	2590	2385	21,661	19,944	5.0	18.821	23.821	3.0	4.0	18.821	3.629	12.074	15.703	2.994	1.634	12.074	0.023
10	n-Butane	C_4H_{10}	58.118	0.1582[e]	6.321[e]	2.06654[e]	3370	3113	21,308	19,680	6.5	24.467	30.967	4.0	5.0	24.467	3.579	11.908	15.487	3.029	1.550	11.908	0.022
11	Isobutane	C_4H_{10}	58.118	0.1582[e]	6.321[e]	2.06654[e]	3363	3105	21,257	19,629	6.5	24.467	30.967	4.0	5.0	24.467	3.579	11.908	15.487	3.029	1.550	11.908	0.019
12	n-Pentane	C_5H_{12}	72.144	0.1904[e]	5.252[e]	2.4872[e]	4016	3709	21,091	19,517	8.0	30.114	38.114	5.0	6.0	30.114	3.548	11.805	15.353	3.050	1.498	11.805	0.025
13	Isopentane	C_5H_{12}	72.144	0.1904[e]	5.252[e]	2.4872[e]	4008	3716	21,052	19,478	8.0	30.114	38.114	5.0	6.0	30.114	3.548	11.805	15.353	3.050	1.498	11.805	0.071
14	Neopentane	C_5H_{12}	72.144	0.1904[e]	5.252[e]	2.4872[e]	3993	3693	20,970	19,396	8.0	30.114	38.114	5.0	6.0	30.114	3.548	11.805	15.353	3.050	1.498	11.805	0.11
15	n-Hexane	C_6H_{14}	86.169	0.2274[e]	4.398[e]	2.9704[e]	4762	4412	20,940	19,403	9.5	35.760	45.260	6.0	7.0	35.760	3.528	11.738	15.266	3.064	1.464	11.738	0.05
	Olefin series C_nH_{2n}																						
16	Ethylene	C_2H_4	28.051	0.07456	13.412	0.9740	1613.8	1513.2	21,644	20,295	3.0	11.293	14.293	2.0	2.0	11.293	3.422	11.385	14.807	3.138	1.285	11.385	0.021
17	Propylene	C_3H_6	42.077	0.1110[e]	9.007[e]	1.4504[e]	2336	2186	21,041	19,691	4.5	16.939	21.439	3.0	3.0	16.939	3.422	11.385	14.807	3.138	1.285	11.385	0.031
18	n-Butene (Butylene)	C_4H_8	56.102	0.1480[e]	6.756[e]	1.9336[e]	3084	2885	20,840	19,496	6.0	22.585	28.585	4.0	4.0	22.585	3.422	11.385	14.807	3.138	1.285	11.385	0.031
19	Isobutene	C_4H_8	56.102	0.1480[e]	6.756[e]	1.9336[e]	3068	2869	20,730	19,382	6.0	22.585	28.585	4.0	4.0	22.585	3.422	11.385	14.807	3.138	1.285	11.385	0.031
20	n-Pentene	C_5H_{10}	70.128	0.1852[e]	5.400[e]	2.4190[e]	3836	3586	20,712	19,363	7.5	28.232	35.732	5.0	5.0	28.232	3.422	11.385	14.807	3.138	1.285	11.385	0.037
	Aromatic series C_nH_{2n-6}																						
21	Benzene	C_6H_6	78.107	0.2060[e]	4.852[e]	2.6920[e]	3751	3601	18,210	17,480	7.5	28.232	35.732	6.0	3.0	28.232	3.073	10.224	13.297	3.381	0.692	10.224	0.12
22	Toluene	C_7H_8	92.132	0.2431[e]	4.113[e]	3.1760[e]	4484	4284	18,440	17,620	9.0	33.878	42.878	7.0	4.0	33.878	3.126	10.401	13.527	3.344	0.782	10.401	0.21
23	Xylene	C_8H_{10}	106.158	0.2803[e]	3.567[e]	3.6618[e]	5230	4980	18,650	17,760	10.5	39.524	50.024	8.0	5.0	39.524	3.165	10.530	13.695	3.317	0.849	10.530	0.36
	Miscellaneous gases																						
24	Acetylene	C_2H_2	26.036	0.06971	14.344	0.9107	1499	1448	21,500	20,776	2.5	9.411	11.911	2.0	1.0	9.411	3.073	10.224	13.297	3.381	0.692	10.224	0.16
25	Naphthalene	$C_{10}H_8$	128.162	0.3384[e]	2.955[e]	4.4208[e]	5854[f]	5654[f]	17,298[f]	16,708[f]	12.0	45.170	57.170	10.0	4.0	45.170	2.996	9.968	12.964	3.434	0.562	9.968	—[f]
26	Methyl alcohol	CH_3OH	32.041	0.0846[e]	11.820[e]	1.1052[e]	867.9	768.0	10,259	9,078	1.5	5.646	7.146	1.0	2.0	5.646	1.498	4.984	6.482	1.374	1.125	4.984	0.027
27	Ethyl alcohol	C_2H_5OH	46.067	0.1216[e]	8.221[e]	1.5890[e]	1600.3	1450.5	13,161	11,929	3.0	11.293	14.293	2.0	3.0	11.293	2.084	6.934	9.018	1.922	1.170	6.934	0.030
28	Ammonia	NH_3	17.031	0.0456[e]	21.914[e]	0.5961[e]	441.1	365.1	9,668	8,001	0.75	2.823	3.573	—	1.5	3.323	1.409	4.688	6.097	—	1.587	5.511	0.088
29	Sulfur	S	32.06	—	—	—	—	—	3,983	3,983	—	—	—	—	—	—	0.998	3.287	4.285	SO_2 1.998	—	3.287	0.071
30	Hydrogen sulfide	H_2S	34.076	0.09109[e]	10.979[e]	1.1898[e]	647	596	7,100	6,545	1.5	5.646	7.146	SO_2 1.0	1.0	5.646	1.409	4.688	6.097	SO_2 1.880	0.529	4.688	0.30
31	Sulfur dioxide	SO_2	64.06	0.1733	5.770	2.264	—	—	—	—	—	—	—	—	—	—	—	—	—	—	—	—	—
32	Water vapor	H_2O	18.016	0.04758[e]	21.017[e]	0.6215[e]	—	—	—	—	—	—	—	—	—	—	—	—	—	—	—	—	—
33	Air	—	28.9	0.07655	13.063	1.0000	—	—	—	—	—	—	—	—	—	—	—	—	—	—	—	—	—

Source: V. Ganapathy, *Steam Plant Calculations Manual*, Marcel Dekker, New York, 1984.

TABLE 1.2C Nomenclature for Program 1.2

Nomenclature	Program symbol	Description and units
a_i	-	Air required (ft^3/ft^3 fuel)
b_i	-	Constant from Table 1.2B
c_i	-	Constant from Table 1.2B
E	E	Excess air (%)
MW_f	MWF	Molecular weight of fuel
MW_g	MWFG	Molecular weight of flue gas
p_c	CO2W	$\%CO_2$ in wet flue gas
p_n	N2W	$\%N_2$ in wet flue gas
p_o	O2W	$\%O_2$ in wet flue gas
p_w	H2O	$\%H_2O$ in wet flue gas
p_s	SO2W	$\%SO_2$ in wet flue gas
—	CO2D	$\%CO_2$ in dry flue gas
—	N2D	$\%N_2$ in dry flue gas
—	O2D	$\%O_2$ in dry flue gas
—	SO2D	$\%SO_2$ in dry flue gas
RH	RH	Relative humidity (fraction)
SVP	SVP	Saturation vapor pressure (psia)
t	T	Ambient temperature (°F)
t_d	TDP	Water dew point (°F)
v_a	—	Moles of dry air per mole fuel
v_g	VFG	Moles of wet flue gas per mole fuel
v_m	—	Moles of water vapor per mole fuel
w_{wa}	WWA	Wet air required (lb/lb fuel)
w_{wg}	WWG	Wet gas produced (lb/lb fuel)
HHV_v	HHV	Higher heating value (Btu/ft^3)
HHV_w	HHVW	Higher heating value (Btu/lb)
LHV_v	LHV	Lower heating value in (Btu/ft^3)
LHV_w	LHVW	Lower heating value (Btu/lb)
ρ_f	—	Fuel density (lb/ft^3 at 60°F)
y_i	A(I)	%Volume of constituent gas

The weight of wet air per pound of fuel w_{wa} is

$$w_{wa} = \frac{18v_m + 29v_t}{MW_f} \tag{18}$$

The wet flue gas per pound of fuel is

$$w_{wg} = v_g \times \frac{MW_g}{MW_f} \tag{19}$$

where MW_g and MW_f are molecular weights of flue gas and fuel, respectively.

$$MW_g = \sum \frac{y_i MW_i}{100} \tag{20}$$

Higher and lower heating values of fuel are obtained as follows:

$$HHV_v = \sum HHV_i \frac{y_i}{100} \tag{21}$$

$$LHV_v = \sum LHV_i y_i \tag{22}$$

$$HHV_w = HHV_v \frac{378.1}{MW_f} \tag{23}$$

$$LHV_w = LHV_v \frac{378.1}{MW_f} \tag{24}$$

where

HHV_v = higher heating value (Btu/ft^3)
LHV_v = lower heating value (Btu/ft^3)
HHV_w = higher heating value (Btu/lb)
LHV_w = lower heating value (Btu/lb)

Example

A gaseous fuel has the analysis: CH_4 = 83.4, C_2H_6 = 15.8, and N_2 = 0.8%, all by volume. Perform detailed combustion calculations.

```
10 DIM AIR(30),MW(30),CO2FG(30),H2OFG(30),HHV(30),LHV(30),A(20)
20 FOR I=1 TO 19
30 READ AIR(I),MW(I),CO2FG(I),H2OFG(I),HHV(I),LHV(I)
40 NEXT I
50 DATA 9.528,16.041,1,2,1013.2,913.1,16.675,30.067,2,3,1792,1641,23.821,44.092,
3,4,2590,2385,30.967,58.118,4,5,3370,3113,38.114,72.144,5,6,4016,3709,14.293,28.
051,2,2,1613.2,1513.2
60 DATA 21.439,42.077,3,3,2336,2186,28.585,56.102,4,4,3084,2885,35.732,78.1,6,3,
3751,3601,42.878,92.132,7,4,4484,4284,11.911,26.036,2,1,1499,1448,3.573,17.031,0
,1.5,441.1,365.1
70 DATA 7.146,34.076,1,1,647,596,0,18.016,0,0,0,0,0,28.01,0,0,0,0,0,44.01,0,0,0,
0,2.382,28.01,1,0,321.8,321.8,2.382,2.016,0,1,325,275,0,64.06,0,0,0,0
80 INPUT"EXCESS AIR %,AMB TEMP,REL HUM-FRACTION=";E,T,RH
90 INPUT"GAS NO,ANALYSIS=";I,A(I)
100 IF A(I)=0 GOTO 120
110 GOTO 90
120 HHV=0:LHV=0:AIR=0:MWF=0:CO2FG=0:H20FG=0
130 FOR I=1 TO 19
140 HHV=HHV+.01*A(I)*HHV(I)
150 LHV=LHV+.01*A(I)*LHV(I)
160 AIR=AIR+.01*A(I)*AIR(I)
170 CO2FG=CO2FG+.01*A(I)*CO2FG(I)
180 H2OFG=H2OFG+.01*A(I)*H2OFG(I)
190 MWF=MWF+.01*A(I)*MW(I)
200 NEXT I
210 HHVW=HHV*378.1/MWF
220 LHVW=LHV*378.1/MWF
230 AIRA=AIR*(1+.01*E):SVP=.08+281*10^-9*T^3.25:VM=SVP*RH/(14.7-SVP*RH)
240 VH2O=H2OFG+VM*AIRA:VN2=.79*AIRA+.01*A(15)
250 VSO2=.01*(A(13)+A(19)):VO2=.0021*AIR*E
260 VFG=VSO2+VO2+VN2+VH2O+CO2FG:CO2W=100*CO2FG/VFG:O2W=100*VO2/VFG
270 H2O=100*VH2O/VFG:N2W=100*VN2/VFG:SO2W=100*VSO2/VFG:F=100/(100-H2O)
280 CO2D=CO2W*F:O2D=O2W*F:N2D=N2W*F:SO2D=SO2W*F
290 WWA=(18*VM*AIRA+29*AIRA)/MWF:MWFG=(CO2W*.44+O2W*.32+H2O*.18+.28*N2W+.64*SO2W
)
295 TDP=((.147*H2O-.08)*10^6/.281)^.3077
300 WWG=WWA+1
310 PRINT" "
320 PRINT"COMBUSTION CALCULATIONS-GASEOUS FUEL"
330 PRINT" "
340 PRINT"FUEL DATA"
345 PRINT"GAS NO   %VOLUME"
350 FOR I=1 TO 19
355 IF A(I)=0 GOTO 370
360 PRINT I,A(I)
```

FIG. 1.2 Listing of program 1.2, with results.

Solution

Key in the program, a listing of which is in Fig. 1.2. In the RUN mode, the screen asks for ambient temperature (70), relative humidity (0.80) and excess air (15), which are fed in. Then, the

```
370 NEXT I
380 PRINT"EXCESS AIR %=";E;"AMB TEMP=";T;"REL HUM-FRACTION=";RH
390 PRINT" "
400 PRINT"MOL WT FUEL=";MWF;"MOL WT FLUE GAS=";MWFG
410 PRINT"HHV -BTU/LB=";HHVW;"LHV -BTU/LB=";LHVW
420 PRINT"HHV-BTU/CU FT=";HHV;"LHV-BTU/CU FT=";LHV
430 PRINT" "
440 PRINT"WET GAS FLUE GAS ANALYSIS-% VOL"
450 PRINT" "
460 PRINT"CO2 =";CO2W;"H2O =";H2O;"O2 =";O2W;"N2 =";N2W;"SO2=";SO2W
470 PRINT" "
480 PRINT"DRY FLUE GAS ANALYSIS-%VOL"
490 PRINT" "
500 PRINT"CO2=";CO2D;"O2 =";O2D;"N2 =";N2D;"SO2=";SO2D
510 PRINT" "
520 PRINT"WET  AIR-LB/LB FUEL=";WWA;"WET FLUE GAS-LB/LB FUEL=";WWG
525 PRINT" "
530 PRINT"DEW POINT OF WATER VAPOR-F=";TDP
535 PRINT" "
540 END
```

```
GAS NO   %VOLUME
 1          83.4
 2          15.8
 15         .8
EXCESS AIR %= 15 AMB TEMP= 70 REL HUM-FRACTION= .8

MOL WT FUEL= 18.35286 MOL WT FLUE GAS= 27.69529
HHV -BTU/LB= 23241.69 LHV -BTU/LB= 21030.28
HHV-BTU/CU FT= 1128.145 LHV-BTU/CU FT= 1020.803

WET GAS FLUE GAS ANALYSIS-% VOL

CO2 = 8.525802 H2O = 17.67677 O2 = 2.471011 N2 = 71.32641 SO2= 0

DRY FLUE GAS ANALYSIS-%VOL

CO2= 10.3565 O2 = 3.001597 N2 = 86.64191 SO2= 0

WET  AIR-LB/LB FUEL= 19.46499 WET FLUE GAS-LB/LB FUEL= 20.46499

DEW POINT OF WATER VAPOR-F= 137.8046
```

FIG. 1.2 (Continued)

gas number and %volume of constituents present are fed in. Then, to execute the program, input the gas number of a gas not present, say, 4, and its volume, 0. The computer then solves all of the equations and prints out the results, as seen in Fig. 1.2.

PROGRAM 1.3 EFFICIENCY OF BOILERS AND HEATERS BASED ON ASME POWER TEST CODE

Input

Carbon
Hydrogen
Oxygen
Sulfur
Moisture
Nitrogen
Excess air
Ambient temperature
Exit gas temperature
Relative humidity
Radiation loss
Duty
Unaccounted loss

Output

Dry flue gas loss
Air moisture loss
Fuel moisture loss
Radiation loss
Efficiency on higher heating value basis
Efficiency on lower heating value basis
Higher heating value of fuel
Lower heating value of fuel

Remarks

This program is applicable for solid and liquid fuels. It uses the ASME PTC 4.1 method of losses. Radiation loss is also determined based on the ABMA curve, and the higher input of radiation loss and ABMA value are used. Radiation loss can also be input as zero, and the program then uses the ABMA value.

The program calculates the various losses associated with the combustion of solid and liquid fuels based on the ASME power test code* and predicts the efficiency on a higher and a lower heating

*ASME power test code, Performance test code for steam generating units, PTC 4.1.

value basis if the exit gas temperature and ambient conditions are known. The effect of relative humidity is considered. The program accepts any desired radiation loss and uses the higher of the ABMA chart value and the input radiation loss.

Theory

Given the fuel ultimate analysis, combustion calculations may be performed as follows.

$$HHV = 14{,}500C + 62{,}000(H_2 - 0.125O_2) + 4000S \qquad (1)$$

$$LHV = HHV - 9720H_2 - 1110W \qquad (2)$$

$$SVP = 0.08 + 0.281 \times 10^{-6} t_a^{3.25} \qquad (3)$$

$$M = \frac{0.00622 \times RH \times SVP}{14.7 - 0.01 \times RH \times SVP} \qquad (4)$$

$$w_{at} = 22.53C + 34.3(H_2 - 0.125O_2) + 4.29S \qquad (5)$$

$$w_a = w_{at}(1 + 0.01E) \qquad (6)$$

$$G = 8.49H_2 + W + 0.01w_a EM \qquad (7)$$

The various losses are as follows.

$$L_1 = [24w_a(1 + M) + (1 - G)]\frac{t_g - t_a}{HHV} \qquad (8)$$

$$L_2 = 100(9H_2 + W)\frac{1080 + 0.46t_g - t_a}{HHV} \qquad (9)$$

$$L_3 = 46Mw_a(1 + 0.01E)\frac{t_g - t_a}{HHV} \qquad (10)$$

$$L_4 = 10^{0.62 - 0.42\log Q} \qquad (11)$$

Equation 11 is based on the ABMA radiation loss chart. However, the program will also accept any radiation loss input

TABLE 1.3 Nomenclature for Program 1.3

Nomenclature	Program symbol	Description and units
C	C	Carbon in fuel (fraction)
E	E	Excess air (%)
G	G	Moisture (lb/lb fuel)
H_2	H	Hydrogen in fuel (fraction)
HHV	V	Higher heating value (Btu/lb)
L_1	I	Dry gas loss (%)
L_2	J	Loss due to fuel moisture (%)
L_3	K	Loss due to air moisture (%)
L_4	N	Radiation loss (%)
L_5	X	Unaccounted loss (%)
LHV	L	Lower heating value (Btu/lb)
M	M	Moisture in air (lb/lb dry air)
N_2	Y	Nitrogen in fuel (fraction)
O_2	O	Oxygen in fuel (fraction)
Q	Q	Duty of boiler (MM Btu/hr)
RH	R	Relative humidity (%)
S	S	Sulfur in fuel (fraction)
SVP	P	Saturated vapor pressure (psia)
t_a	B	Ambient temperature (°F)
t_g	T	Exit gas temperature (°F)
w_a	—	Actual air (lb/lb fuel)
w_{at}	A	Theoretical air (lb/lb fuel)
η_{HHV}	A27	Efficiency, HHV basis (%)
η_{LHV}	A28	Efficiency, LHV basis (%)

and use the larger of the two. This provision is given because several heat recovery boiler and fired heater manufacturers use a value for radiation loss different from the value obtained from the ABMA chart, which is widely used by utility and industrial boiler designers. For the nomenclature, see Table 1.3.

Then, efficiency based on higher heating value is given by

$$\eta_{HHV} = 100 - (L_1 + L_2 + L_3 + L_4 + L_5) \tag{12}$$

where L_5 is the unaccounted loss or margin, which the boiler designer would use in the evaluation of efficiency.

Once the efficiency on a HHV basis is obtained, it can be converted to an LHV basis using the equation

$$\eta_{LHV} = \eta_{HHV} \frac{HHV}{LHV} \tag{13}$$

Figure 1.3 gives the listing of the program.

Example

An oil-fired boiler of duty 150 MM Btu/hr has the following data (fuel data are in fractions by weight):

Carbon = 0.865
Hydrogen = 0.125
Oxygen = 0
Sulfur = 0.01
Moisture = 0
Nitrogen = 0
Excess air = 20%
Ambient temperature = 80°F
Exit gas temperature = 350°F
Relative humidity = 60%
Radiation loss = 1.5% (if unknown, input 0 and the program uses the ABMA value)
Duty = 150
Unaccounted loss = 0%

Determine the efficiency based on a higher and lower heating value of fuel and the various losses.

Solution

Key in the program. In the RUN mode, the screen display asks for the data in the same order given here. Once these are fed in, the program solves the equations and prints out the results shown in Fig. 1.3. It may be seen that the efficiency on a higher heating value basis is 86.4%; that on a lower heating value basis is 91.9%. Heating values and the losses are also printed out.

```
10 PRINT"BOILER EFFICIENCY USING ASME PTC METHOD -SOLID/LIQUID FUELS"
20 INPUT"CARBON,HYDROGEN,OXYGEN,SULFUR,MOISTURE,NITROGEN-FRACTIOS=";C,H,O,S,W,Y
30 INPUT"EXCESS AIR %,AMBIENT TEMP,EXIT GAS TEMP,REL HUM %=";E,B,T,R
40 INPUT"RADIATION LOSS %-PROGRAM COMPUTES THE LARGER OF THIS VALUE AND THE ABMA
 VALUE -IF ABMA VALUE IS TO BE USED INPUT ZERO=";A29
50 INPUT"DUTY MM BTU/H(HHV BASIS),UNACCOUNTED LOSS %=";Q,X
60 V=14500*C+62000l*(H-.125*O)+4000*S:L=V-9720*H-1110*W
70 P=.08+281*10^-9*B^3.25:M=.00622*R*P/(14.7-.01*R*P)
80 A=11.53*C+34.3*(H-.125*O)+4.29*S:G=8.939999*H+.01*A*E*M+W
90 I=24*(A*(1+.01*E)*(1+M)+1-G)*(T-B)/V:J=100*(9*H+W)*(1080+.46*T-B)/V
100 K=46*M*A*(1+.01*E)*(T-B)/V:N=10^(.62-.1824*LOG(Q)):N=.5*(A29+N+ABS(A29-N))
110 A27=100-(I+J+K+N+X):A28=A27*V/L
115 PRINT" "
120 PRINT"RESULTS-BOILER EFFICIENCY"
130 PRINT" "
140 PRINT"FUEL DATA"
150 PRINT" "
160 PRINT"CARBON=";C;"HYGROGEN=";H;"MOISTURE=";W;"OXYGEN=";O
170 PRINT" "
180 PRINT"SULFUR=";S;"NITROGEN=";Y
190 PRINT" "
200 PRINT"AMB TEMP=";B;"EXIT GAS TEMP=";T;"REL HUM %=";R;"EXCESS AIR %=";E
210 PRINT" "
220 PRINT"FUEL HHV-BTU/LB=";V;"FUEL LHV-BTU/LB=";L
230 PRINT" "
240 PRINT"BOILER HEAT LOSSES-%"
250 PRINT" "
260 PRINT"DRY GAS LOSS="I;"AIR MOIST LOSS=";K;"FUEL MOIS LOSS=";J
270 PRINT" "
280 PRINT"RAD LOSS=";N;"UNACC LOSS=";X
290 PRINT" "
300 PRINT"EFFICIENCY  HHV BASIS %=";A27;"LHV BASIS %=";A28
305 PRINT" "
310 END

DUTY MM BTU/H(HHV BASIS),UNACCOUNTED LOSS %=? 150,0

RESULTS-BOILER EFFICIENCY

FUEL DATA

CARBON= .865 HYGROGEN= .125 MOISTURE= 0 OXYGEN= 0

SULFUR= .01 NITROGEN= 0

AMB TEMP= 80 EXIT GAS TEMP= 350 REL HUM %= 60 EXCESS AIR %= 20

FUEL HHV-BTU/LB= 20332.5 FUEL LHV-BTU/LB= 19117.5

BOILER HEAT LOSSES-%

DRY GAS LOSS= 5.493257 AIR MOIST LOSS= .1387193 FUEL MOIS LOSS= 6.423829

RAD LOSS= 1.5 UNACC LOSS= 0

EFFICIENCY  HHV BASIS %= 86.4442 LHV BASIS %= 91.9381
```

FIG. 1.3 Listing of program 1.3, with results.

PROGRAM 1.4 EFFICIENCY OF BOILERS AND HEATERS BASED ON FIELD DATA

Input

Fuel type (coal, oil, or gas)
Higher heating value
Lower heating value
%Oxygen, nitrogen, and carbon monoxide in dry flue gas by volume
Exit gas temperature
Ambient temperature
Radiation and unaccounted losses

Output

Excess air
Wet flue gas loss
Efficiency on higher heating value basis
Efficiency on lower heating value basis

Remarks

This method may be used for good engineering estimates of boiler efficiency in the absence of fuel ultimate analysis.

The program calculates the efficiency of boilers and heaters fired on coal, oil, or gas on higher and lower heating value basis if the flue analysis, exit gas, and ambient temperatures are known. Fuel ultimate analysis is not needed. The estimate is adequate for engineering purposes.

Theory

Excess air from the dry flue flue gas analysis is obtained from*

$$E = \frac{100(O_2 - CO/2)}{0.264N_2 - (O_2 - CO/2)} \tag{1}$$

---

*V. Ganapathy, *Applied Heat Transfer*, Pennwell Books, Tulsa, Oklahoma, 1982, p. 22.

The wet flue gas quantity is given by

$$w = \left[A(1 + 0.01E) + \frac{10^6}{HHV}\right]\frac{HHV}{10^6} \qquad (2)$$

The wet flue gas loss, on LHV basis, is given by

$$L = 24w\,\frac{t_g - t_a}{LHV} \qquad (3)$$

If Z is the radiation and unaccounted loss (%), the efficiency on a LHV basis is given by

$$\eta_{LHV} = 100 - L - Z \qquad (4)$$

Then, efficiency on a HHV basis is obtained from

$$\eta_{HHV} = \eta_{LHV}\,\frac{LHV}{HHV} \qquad (5)$$

Table 1.4 shows the nomenclature. An example illustrates the use of the program, a listing of which is shown in Fig. 1.4.

Example

Flue gas analysis on a dry basis performed on an oil-fired boiler gives the following data.

$\%O_2 = 3.2$
$\%N_2 = 78$
%CO = nil
Exit gas temperature = 350°F
Ambient temperature = 80°F

If fuel higher and lower heating values are 18,450 and 17,345 Btu/lb, respectively, determine the excess air, flue gas quantity, flue gas loss, and efficiencies on a higher and a lower heating value basis. Assume that radiation and unaccounted losses total 2%.

Solution

Key in the program, a listing of which is given in Fig. 1.4. The display asks for the data in the same order as given here:

TABLE 1.4 Nomenclature

Nomenclature	Program symbol	Description and units
A	A	Factor used in combustion calculations (lb/MM Btu)
E	E	Excess air (%)
CO	C	%Carbon monoxide in dry flue gas by volume
L	X	Wet flue gas loss (%)
N_2	N	%Nitrogen in dry flue gas
O_2	O	%Oxygen in dry flue gas
HHV	H	Higher heating value (Btu/lb)
LHV	L	Lower heating value (Btu/lb)
t_a	B	Ambient temperature (°F)
t_g	T	Exit gas temperature (°F)
w	W	Wet flue gas (lb/lb fuel)
Z	Z	Radiation and unaccounted losses (%)
η_{HHV}	J	Efficiency on HHV basis (%)
η_{LHV}	I	Efficiency on LHV basis (%)

Fuel number (2) (input 1 for coal, 2 for oil, and 3 for natural gas)
HHV (18,450)
LHV (17,345)
$\%O_2$ (3.2)
$\%N_2$ (78)
%CO (0)
Exit gas temperature (350)
Ambient temperature (80)
Radiation and unaccounted losses (2)

Once these data are input, the program solves the various equations and prints out the results. It is seen that excess air is 18%, flue gas loss = 6.45%, efficiency on LHV basis is 91.54%.

```
10 PRINT"EFFICIENCY OF BOILERS FROM FIELD DATA"
20 INPUT"FUEL TYPE NO-1 FOR COAL,2 FOR OIL,3 FOR N.GAS=";K
30 INPUT"FUEL HIGHER HEATING VALUE,LOWER HEATING VALUE-BTU/LB=";H,L
40 INPUT"DRY FLUE GAS ANALYSIS%-CO,O2,N2=";C,O,N
50 INPUT"EXIT GAS TEMP,AMB TEMP,RADIATION PLUS UNACC LOSSES=";T,B,Z
60 E=(O-.5*C)/(.264*N-(O-.5*C))
70 A=ABS((K=1)*760+(K=2)*745+(K=3)*730 )
80 W=((A*(1+E)+10^6/H)*H/10^6):X=24*(T-B)*W/L
90 I=100-X-Z:E=100*E:J=I*L/H
95 PRINT" "
100 PRINT"RESULTS"
110 PRINT" "
120 PRINT"FUEL TYPE-1 IS COAL,2 IS OIL,3 IS N.GAS=";K
130 PRINT" "
140 PRINT"HHV-BTU/LB=";H;"LHV-BTU/LB=";L
150 PRINT" "
160 PRINT"EXCESS AIR=";E
165 PRINT" "
170 PRINT"AMB TEMP=";B;"EXIT TEMP=";T
180 PRINT" "
190 PRINT"FLUE GAS-LB/LB FUEL=";W;"%CO=";C;"%N2=";N;"%O2=";O
195 PRINT" "
200 PRINT"FLUE GAS LOSS=";X;"OTHER LOSSES=";Z;"EFF ON LHV BASIS-%=";I
210 PRINT" "
220 PRINT"EFF ON HHV BASIS-%=";J
230 END

RUN
EFFICIENCY OF BOILERS FROM FIELD DATA
FUEL TYPE NO-1 FOR COAL,2 FOR OIL,3 FOR N.GAS=? 2
FUEL HIGHER HEATING VALUE,LOWER HEATING VALUE-BTU/LB=? 18450,17345
DRY FLUE GAS ANALYSIS%-CO,O2,N2=? 0,3.2,78
EXIT GAS TEMP,AMB TEMP,RADIATION PLUS UNACC LOSSES=? 350,80,2

RESULTS

FUEL TYPE-1 IS COAL,2 IS OIL,3 IS N.GAS= 2

HHV-BTU/LB= 18450 LHV-BTU/LB= 17345

EXCESS AIR= 18.39927

AMB TEMP= 80 EXIT TEMP= 350

FLUE GAS-LB/LB FUEL= 17.27428 %CO= 0 %N2= 78 %O2= 3.2

FLUE GAS LOSS= 6.453578 OTHER LOSSES= 2 EFF ON LHV BASIS-%= 91.54642

EFF ON HHV BASIS-%= 86.06356
```

FIG. 1.4 Listing of program 1.4, with results.

2

Fluid Flow and Pressure Drop Calculations

PROGRAM 2.1 PRESSURE DROP OF SATURATED AND SUPERHEATED STEAM IN PIPES

Input

Steam pressure
Temperature (if saturated, input 0 will enable program to estimate saturation temperature)
Flow
Pipe inner diameter

Output

Steam Temperature
Velocity
Specific volume
Pressure drop per 100 ft of pipe

Remarks

ASME 1967 formulations are used to compute steam properties. Saturated steam temperature, if required, is computed by the program by the input of 0 for steam temperature.

The program calculates the pressure drop of saturated and superheated steam in tubes and pipes for a length of 100 ft, given the steam pressure and temperature, flow, and pipe inner diameter. If the steam is in a saturated condition, the input of zero for temperature will accomplish the calculation of the saturation temperature. Ther is no need to refer to the steam tables for the properties. The program also computes the steam velocity and prints out the pressure drop, specific volume, pressure, and temperature, and velocity. ASME 1967 formulations are used in the estimation of steam properties. The Appendix gives these correlations.

Theory

Pressure drop of fluids flowing in pipes or tubes may be estimated from the following equations (for a length of 100 ft of pipe)*:

$$\Delta P = \frac{3.36 \times 10^{-4} f W^2 v}{d_i^5} \qquad (1)$$

*V. Ganapathy, *Applied Heat Transfer*, Pennwell Books, Tulsa, Oklahoma, 1982, p. 522.

$$f = 0.0055 \left\{ 1 + \left[\left(\frac{36}{d_i} \right) + 1 \right]^{0.33} \right\} \tag{2}$$

The velocity V is given by

$$V = \frac{.05Wv}{d_i^{\,2}} \tag{3}$$

Table 2.1 gives the nomenclature and Fig. 2.1 the listing.

If the steam is in a saturated condition, the input of zero for the temperature results in estimation of the saturation temperature using the equation

$$t = 115P^{0.225} \tag{4}$$

Steam temperature and pressure are required to estimate specific volume.

Example 1

In a boiler superheater, steam conditions are as follows:

Pressure = 1000 psia
Temperature = saturated (use 0 if saturation temperature is unknown)
Flow = 4500 lb/hr
Tube inner diameter = 1.75 in.

TABLE 2.1 Nomenclature for Progam 2.1

Nomenclature	Program symbol	Description and units
d_i	D	Pipe or tube inner diameter (in.)
f	F	Friction factor
P	P	Steam pressure (psia)
ΔP	Z	Pressure drop per 110 ft (psi)
t	T	Steam temperature (°F)
v	V	Specific volume (ft^3/lb)
V	S	Velocity (ft/sec)
W	WS	Steam flow (lb/hr)

```
10 PRINT"PRESSURE DROP OF STEAM IN PIPES"
20 INPUT"PRESSURE-PSIA,TEMPERATURE(IF SATURATED IN PUT 0 AND PROGRAM COMPUTES TE
MP),FLOW-LB/H,PIPE ID-IN=";P,T,WS,D
30 IF T=0 THEN T=115*P^.225
40 GOSUB 500
50 S=.05*WS*V/D/D:F=.0055*(1+((36/D)+1)^.33)
60 Z=.000336*WS*WS*V*F/D^5
70 PRINT" "
80 PRINT"PRESSURE DROP OF STEAM"
90 PRINT" "
100 PRINT"FLOW=";WS;"TEMP=";T;"PRESS-PSIA=";P
110 PRINT" "
120 PRINT"SP VOL=";V;"VELOCITY-FT/S=";S;"PRESS DROP/100 FT-PSI=";Z
130 PRINT" "
140 PRINT"PIPE ID-IN=";D
150 END
500 T=273.1+(T-32)/1.8:P=P/14.696
510 K=808701/T/T:L=10^K*(-2641.62/T):M=1.89+L:N=M*P*P/T/T
520 O=2+(3724201/T/T):Q=O*L:R=1.89+Q:U=(.21828*T-1269701/T):V=2*U*R-(M/T)*126460
!
530 W=82.54-1624601/T:Y=2*W*R-(M/T)*1624601
540 V=(((U*M*N+W)*N/P+1)*M+4.5504*T/P)*.0160185
590 P=14.696*P:T=(T-273.15)*1.8+32
600 RETURN
```

```
PRESSURE DROP OF STEAM

FLOW= 4500 TEMP= 544.0339 PRESS-PSIA= 999.9999

SP VOL= .4449299 VELOCITY-FT/S= 32.68873 PRESS DROP/100 FT-PSI= 3.809615

PIPE ID-IN= 1.75


PRESSURE DROP OF STEAM IN PIPES
PRESSURE-PSIA,TEMPERATURE(IF SATURATED IN PUT 0 AND PROGRAM COMPUTES TEMP),FLOW-
LB/H,PIPE ID-IN=? 1000,650,4500,1.75

PRESSURE DROP OF STEAM

FLOW= 4500 TEMP= 649.91 PRESS-PSIA= 999.9999

SP VOL= .5633157 VELOCITY-FT/S= 41.38646 PRESS DROP/100 FT-PSI= 4.823268

PIPE ID-IN= 1.75
```

FIG. 2.1 Listing of program 2.1, with results.

Determine the velocity, specific volume, saturation temperature, and pressure drop per 100 ft of tube.

Solution

Key in the program. In the RUN mode, the screen asks for the data in the same order given; they are fed in. The program solves all the equations and prints out the results. It is seen from Fig. 2.1 that specific volume is 0.445 ft^3/lb, velocity is 32 ft/sec, and pressure drop per 100 ft is 3.81 psi. Saturation temperature is 544°F.

Example 2

Use superheated steam at 650°F at the same pressure and flow conditions.

Solution

It is seen that specific volume is 0.564 ft^3/lb and pressure drop is 4.83 psi. Steam velocity is 41 ft/sec.

PROGRAM 2.2 PRESSURE DROP OF WATER IN PIPES

Input

Water flow (either lb/hr or gpm)
Temperature
Pipe inner diameter

Output

Flow (gpm)
Flow (lb/hr)
Water velocity
Specific volume
Pressure drop per 100 ft of pipe

Remarks

The flow rate can be input in lb/hr or in gpm; the program computes both the values.

The program calculates the pressure drop of water in pipes and tubes given the flow (lb/hr or gpm), tube inner diameter, and temperature. Final results include the flow (in both lb/hr and gpm), velocity, and pressure drop per 100 ft of pipe.

Theory

The pressure drop of fluid per 100 ft of pipe is given by

$$\Delta P = \frac{3.36 \times 10^{-4} fW^2 v}{d_i^{\,5}} \tag{1}$$

The friction factor f is obtained from

$$f = 0.0055 \left\{1 + \left[\left(\frac{36}{d_i}\right) + 1\right]^{0.33}\right\} \tag{2}$$

The specific volume v is obtained from

$$\begin{aligned} v &= 0.016 && \text{for } t \leqslant 120^\circ F \\ v &= 0.0149 \exp(0.000555t) && \text{for } t > 120^\circ F \end{aligned} \tag{3}$$

Then the velocity of water V is

$$V = \frac{0.05Wv}{d_i^{\,2}} \tag{4}$$

The flow (lb/hr) is related to gpm as

$$q = 0.125Wv \tag{5}$$

Table 2.2 shows the nomenclature; Fig. 2.2 gives the listing and printout of results.

Example 1

In a boiler plant, 900 gpm of water at 175°F flows in a pipe of inner diameter 5.762 in. Determine the flow (lb/hr), velocity, and pressure drop.

Solution

Key in the program. In the RUN mode, the screen asks for the following data in the same order given:

Water flow (lb/hr = 0, if unknown, input 0)
Flow gpm = 900

TABLE 2.2 Nomenclature for Program 2.2

Nomenclature	Program symbol	Description and units
d_i	D	Pipe or tube inner diameter (in.)
f	F	Friction factor
ΔP	P	Pressure drop per 100 f (psi)
q	Q	Flow in (gpm)
t	T	Water temperature (°F)
v	V	Specific volume (ft^3/lb)
V	Z	Velocity (ft/sec)
W	W	Flow (lb/hr)

Temperature = 175
Pipe ID = 5.762

Once these data are fed in, the program solves the equations and prints out the results. It is seen that flow (lb/hr) = 438,495, velocity is 10.85 ft/sec, and pressure drop is 2.68 psi per 100 ft.

Example 2

A total of 50,000 lb/hr of water at 220°F flows in a pipe of ID = 2.9 in. Determine the velocity, pressure drop, and flow (gpm).

Solution

In the RUN mode the screen asks for the same data as before. We input 50,000 for flow in lb/hr and 0 for flow in gpm. The results are shown in Fig. 2.2: gpm is 105, and pressure drop is 1.27 psi per 100 ft.

PROGRAM 2.3 PRESSURE DROP OF AIR AND FLUE GAS IN TUBES AND PIPES

Input

Flow (lb/hr or scfm)
Pressure
Temperature
Pipe inner diameter

```
10 PRINT"PRESSURE DROP OF WATER IN PIPES"
15 PRINT"FLOW MAY BE GIVEN EITHER IN LB/H OR IN GPM.PROGRAM COMPUTES BOTH.IF FLO
W IS NOT KNOWN IN LB/H OR IN GPM,INPUT ZERO FOR THAT VALUE"
20 INPUT"WATER FLOW-LB/H,WATER FLOW-GPM,TEMPERATURE,PIPE ID-IN=";W,Q,T,D
30 IF T<=120 THEN V=.016
40 IF T>120 THEN V=.0149*EXP(.000555*T)
50 IF W=0 THEN W=8*Q/V
60 IF Q=0 THEN Q=.125*W*V
70 F=.0055*(1+((36/D)+1)^.33):P=.000336*F*W*W*V/D^5
80 Z=.05*W*V/D/D
90 PRINT" "
100 PRINT"PRESSURE DROP-WATER"
110 PRINT" "
120 PRINT"FLOW-LB/H=";W;"FLOW-GPM=";Q;"TEMP=";T
130 PRINT" "
140 PRINT"WATER VELOCITY-FT/S=";Z;"SP VOL-CU FT/LB=";V
150 PRINT" "
160 PRINT"PRESSURE DROP/100 FT-PSI=";P
170 END


PRESSURE DROP-WATER

FLOW-LB/H= 438495.7 FLOW-GPM= 900 TEMP= 175

WATER VELOCITY-FT/S= 10.84316 SP VOL-CU FT/LB= 1.641977E-02

PRESSURE DROP/100 FT-PSI= 2.684661


PRESSURE DROP OF WATER IN PIPES
FLOW MAY BE GIVEN EITHER IN LB/H OR IN GPM.PROGRAM COMPUTES BOTH.IF FLOW IS NOT
KNOWN IN LB/H OR IN GPM,INPUT ZERO FOR THAT VALUE
WATER FLOW-LB/H,WATER FLOW-GPM,TEMPERATURE,PIPE ID-IN=? 50000,0,220,2.9

PRESSURE DROP-WATER

FLOW-LB/H= 50000 FLOW-GPM= 105.2189 TEMP= 220

WATER VELOCITY-FT/S= 5.004464 SP VOL-CU FT/LB= 1.683502E-02

PRESSURE DROP/100 FT-PSI= 1.272416
```

FIG. 2.2 Listing of program 2.2, with results.

Output

Flow (lb/hr)
Flow (acfm)
Flow (scfm)
Velocity

Specific volume
Density
Pressure drop per 100 ft of pipe (psi)
Pressure drop (in. of water column)

Remarks

Flow is calculated in Lb/h, acfm (actual cubic feet per minute), and scfm (standard cubic feet per minute) given the flow either in lb/hr or scfm.

The program calculates the pressure drop of air and flue gases (having nearly same molecular weight as air) inside circular ducts, pipes, or tubes. The effects of pressure and temperature are considered. Flow can be entered either in lb/hr or in scfm (standard cubic feet per minute) and program computes the flow in lb/hr, acfm, and scfm, along with velocity, density, specific volume, and pressure drop per 100 ft (both in psi and in in. of water).

Theory

The following equations are used. Table 2.3 gives the nomenclature.

$$\Delta P = \frac{3.36 \times 10^{-4} f W^2 v}{d_i^5} \tag{1}$$

$$v = \frac{(460 + t)}{2.7P} = \frac{1}{\rho} \tag{2}$$

$$f = 0.0055 \left\{ 1 + \left[\left(\frac{36}{d_i} \right) + 1 \right]^{0.33} \right\} \tag{3}$$

The pressure drop (psi) is related to loss in in. WC as

$$\Delta P = 27.7H \tag{4}$$

The flows (scfm, acfm, and lb/hr) are related as

$$q_s = \frac{W}{4.5} \tag{5a}$$

$$q_a = \frac{Wv}{60} \tag{5b}$$

TABLE 2.3 Nomenclature for Program 2.3

Nomenclature	Program symbol	Description and units
d_i	D	Pipe or tube inner diameter (in.)
f	F	Friction factor
H	H	Pressure drop per 100 ft (in. WC)
P	P	Pressure (psia)
q_a	Q	Flow (acfm)
q_s	N	Flow (scfm)
t	T	Temperature (°F)
v	V	Specific volume (ft^3/lb)
V	S	Velocity (ft/sec)
W	W	Flow (lb/hr)
ΔP	Z	Pressure drop (psi)
ρ	G	Density (lb/ft^3)

The velocity V is given by

$$V = \frac{0.05Wv}{d_i^2} \tag{6}$$

where the scfm is estimated at 70°F. A listing of the program is shown in Fig. 2.3 along with the results.

Example

Air flows in a pipe with the following parameters:

Flow (lb/hr) - 900, scfm = 0; if unknown, input 0
Pressure = 15 psia
Temperature = 60°F
Pipe ID = 3.068 in.

Determine the pressure drop and related parameters.

Solution

Key in the program. In the RUN mode, the screen asks for the data in the same order given; they are fed in. The computer solves

```
10 PRINT"GAS PRESSURE DROP IN PIPES"
15 PRINT"FLOW MAY BE GIVEN EITHER IN LB/H OR IN SCFM AND PROGRAM COMPUTES BOTH T
HE VALUES;INPUT ZERO FOR THE UNKNOWN"
20 INPUT"FLOW-LB/H,FLOW-SCFM,PRESS-PSIA,TEMP-F,PIPE ID=";W,N,P,T,D
30 IF W=0 THEN W=4.5*N
40 IF N=0 THEN N=W/4.5
50 M=.0619+.000042*(T-500):V=(460+T)/2.7/P:G=1/V
60 F=.0055*(1+((36/D)+1)^.33)
70 Z=.000336*F*W*W*V/D^5:H=27.7*Z:S=.05*W*V/D/D:Q=W*V/60
80 PRINT" "
90 PRINT"PRESSURE DROP CALCULATIONS"
100 PRINT" "
110 PRINT"GAS FLOW-LB/H=";W;"FLOW-SCFM=";N;"FLOW-ACFM=";Q
120 PRINT" "
130 PRINT"GAS PRESS-PSIA=";P;"TEMP=";T;"VELOCITY-FT/S=";S
140 PRINT" "
150 PRINT"DENSITY-LB/CU FT=";G;"TUBE ID=";D
160 PRINT" "
170 PRINT"PR DROP-PSI=";Z;"PR DROP-IN WC=";H
180 PRINT" "
190 END
```

```
GAS PRESSURE DROP IN PIPES
FLOW MAY BE GIVEN EITHER IN LB/H OR IN SCFM AND PROGRAM COMPUTES BOTH THE VALUES
;INPUT ZERO FOR THE UNKNOWN
FLOW-LB/H,FLOW-SCFM,PRESS-PSIA,TEMP-F,PIPE ID=? 900,0,15,60,3.068

PRESSURE DROP CALCULATIONS

GAS FLOW-LB/H= 900 FLOW-SCFM= 200 FLOW-ACFM= 192.5926

GAS PRESS-PSIA= 15 TEMP= 60 VELOCITY-FT/S= 61.38328

DENSITY-LB/CU FT= 7.788461E-02 TUBE ID= 3.068

PR DROP-PSI= .2344246 PR DROP-IN WC= 6.493562
```

FIG. 2.3 Listing of program 2.3, with results.

all the equations and prints out the results. It is seen that velocity is 61 ft/sec, pressure drop is 0.23 psi or 6.49 in. WC, flow is 192 acfm (actual cubic feet per minute), or 200 scfm, and density is 0.078 lb/ft^3.

PROGRAM 2.4 FUEL OIL PROPERTIES AND PRESSURE DROP OF OIL IN PIPES

Input

Temperature and viscosity at any two temperatures of oil
Operating temperature
Degree API
Flow (lb/hr)
Pipe inner diameter

Output

Flow (gpm)
Reynolds number
Viscosity of oil (lb/ft hr)
Specific gravity
Density
Pressure drop per 100 ft

This program calculates the properties of fuel oil, such as viscosity, density, and specific gravity at any given temperature and also the pressure drop in lines given the flow and pipe inner diameter. A check is made for the type of flow, that is, laminar or turbulent.

Theory

The pressure drop per 100 ft of pipe is given by the equation

$$\Delta P = \frac{3.36 \times 10^{-4} fW^2}{d_i^5 \rho} \tag{1}$$

where ρ is the density at the operating temperature 5. ρ is obtained from

$$v = \frac{1}{\rho} = \frac{1}{\rho_{60}} [1 + e(t - 60)] \tag{2}$$

ρ_{60} is the density at 60°F, and e is the expansion factor,* which is

$e = 0.0035$ for degree API < 15

$e = 0.00040$ for $15 < \text{API} < 35$

and

$$e = 0.00050 \qquad \text{for API} > 35 \tag{3}$$

The density at 60°F is given by

$$\rho_{60} = \frac{141.5 \times 62.4}{131.5 + \text{API}} \tag{4}$$

The friction factor f depends on whether the flow is laminar or turbulent, which can be determined if the Reynolds number Re is known.

$$\text{Re} = \frac{15.2\text{W}}{d_i \mu} \tag{5}$$

$$f = 64/\text{Re} \qquad \text{for Re} \leqslant 2100 \tag{6a}$$

$$f = 0.0055 \left[1 + \left(\frac{36}{d_i} + \frac{10^6}{\text{Re}} \right)^{033} \right] \qquad \text{for Re} > 2100 \tag{6b}$$

Fuel oil viscosity is available in the form of charts in several readily-available handbooks, including the *North American Handbook* (pp. 20-23).

To arrive at the viscosity at the operating temperature, we need to input the viscosity at any two temperatures. The program then uses the following equation that fits the viscosity curves

$$\log \log \mu = X + Yt \tag{7}$$

The advantage of this approach is that once the viscosity data are fed in at any two temperatures, we can determine the fuel oil properties just by changing the operating temperature. Fig. 2.4 shows the listing of the program, and Table 2.4 gives the nomenclature used.

Example

Fuel oil with the following characteristics flows through the pipe:

```
10 PRINT"PRESSURE DROP IN FUEL OIL LINES"
15 PRINT"FUEL OIL VISCOSITY AT ANY TWO TEMPERATURES"
20 INPUT"TEMP 1,VISCOS 1-CS,TEMP 2,VISC 2=";M,A,N,B
30 INPUT"OPERATING TEMP=";T
40 INPUT"DEGREE API,FLOW-LB/H,PIPE ID=";X,W,I
50 M=273+.5555*(M-32):N=273+.5555*(N-32):T=273+.5555*(T-32)
60 F=(LOG(LOG(A)/2.302))/2.302:G=(LOG(LOG(B)/2.302))/2.302
70 V=10^(10^(F-(F-G)*(LOG(M)-LOG(T))/(LOG(M)-LOG(N))))
80 IF X<15 THEN E=.00035:GOTO 110
90 IF X>35 THEN E=.0005:GOTO 110
100 E=.0004
110 S=((131.5+X)/8830)*(1+E*(1.8*(T-273)-28))
120 D=1/S:V=.03878*V*D:Q=.1246*W/D:R=15.2*W/I/V
130 IF R<2100 THEN FF=64/R:GOTO 150
140 FF=.0055*(1+((36/I)+(10^6/R))^.33)
150 Y=D/62.4:P=3.36*10^(-4)*W*W*FF/D/I^5:T=1.8*(T-273)+32
155 PRINT" "
160 PRINT"RESULTS"
165 PRINT" "
170 PRINT"OIL FLOW-LB/H=";W;"FLOW-GPM=";Q;"OIL TEMP=";T
180 PRINT" "
190 PRINT"PIPE ID=";I;"REY NO=";R;"DEG API-60F=";X
200 PRINT" "
210 PRINT"OIL VISC-LB/FTH=";V;"DENSITY-LB/CU FT=";D;"SP GRAVITY=";Y
220 PRINT" "
230 PRINT"PR DROP/100 FT-PSI=";P;"FRICTION FACTOR=";FF
235 PRINT" "
240 END

PRESSURE DROP IN FUEL OIL LINES
FUEL OIL VISCOSITY AT ANY TWO TEMPERATURES
TEMP 1,VISCOS 1-CS,TEMP 2,VISC 2=? 100,240,140,66
OPERATING TEMP=? 210
DEGREE API,FLOW-LB/H,PIPE ID=? 16,3400,1.049

RESULTS

OIL FLOW-LB/H= 3400 FLOW-GPM= 7.501208 OIL TEMP= 209.9822

PIPE ID= 1.049 REY NO= 1470.911 DEG API-60F= 16

OIL VISC-LB/FTH= 33.49351 DENSITY-LB/CU FT= 56.47624 SP GRAVITY= .9050679

PR DROP/100 FT-PSI= 2.355849 FRICTION FACTOR= 4.351046E-02
```

FIG. 2.4 Listing of program 2.4, with results.

1. At 100°F, μ = 240 cs; at 140°F, μ = 66 cs
2. Operating temperature 210°F
3. Degree API at 60°F = 16.4
4. Flow = 3400 lb/hr
5. Pipe inner diameter = 1.049 in.

TABLE 2.4 Nomenclature for Program 2.4

Nomenclature	Program symbol	Description and units
d_i	I	Pipe inner diameter (in.)
e	E	Expansion factor
f	F	Friction factor
ρ	D	Density (lb/ft^3)
P	P	Pressure drop per 100 ft (psi)
q	Q	Flow (gpm)
Re	R	Reynolds number
s	Y	Specific gravity
t	T	Operating temperature (°F)
v	S	Specific volume (ft^3/lb)
W	W	Flow (lb/hr)
μ	V	Viscosity (lb/ft hr)
API	X	Degree API at 60°F

Calculate the viscosity, density, Reynolds number, flow (gpm), and pressure drop per 100 ft at the operating temperature.

Conversion from lb/hr to gpm is made using the equation

$$q = \frac{0.12467W}{\rho} \tag{8}$$

Solution

Key in the program. In the RUN mode, the screen asks for the data in the same order given; they are fed in. Then, the computer goes on to solve the equations and prints out the results as seen in Fig. 2.4. It is seen that the density is 56.47 lb/ft^3, viscosity at 210°F is 33.46 lb/ft hr, Reynolds number is 1470, and pressure drop per 100 ft is 2.35 psi.

PROGRAM 2.5 SIZING ORIFICES FOR STEAM FLOW

Input

Steam flow
Pressure
Temperature (if saturated, input 0 and program computes saturation temperature)
Pipe inner diameter
Differential head across orifice

Output

Steam temperature
Specific volume
Orifice diameter

Remarks

If saturation temperature is not known, the input of 0 enables the program to compute the value.

The program computes the orifice size given the steam pressure, temperature, pipe size, and differential head. For saturated steam the temperature may or may not be input. A quick convergence logic is used to arrive at the orifice size.

Theory

The basic equation that relates the orifice size with flow* is

$$M = K\beta^2 = \frac{W}{2837 d_i^{\,2} (0.016h/v)^{0.5}} \tag{1}$$

where

$$\beta = \frac{d}{d_i} \tag{2}$$

and

$$K = 0.593 + 0.4\beta^4 + (0.0015\sqrt{\beta} + 0.012\beta^4)\,\frac{1000}{\sqrt{Re}} \tag{3}$$

*Fischer and Porter, *Handbook of Flow Meter Orifice Sizing*, No. 10B9000, Warminster, Pennsylvania.

$$Re = \frac{15.2W}{d_i \mu} \tag{4}$$

The viscosity of steam may be estimated as

$$\mu = 0.016 + 0.000058t \tag{5}$$

The specific volume v of steam is computed from ASME 1967 equations as described in the Appendix. If steam is in a saturated condition, the input of zero for the temperature will result in estimation of t using the equation

$$t = 115P^{0.225} \tag{6}$$

Once M is computed from equation (1), using a trial-and-error procedure, the orifice size d_i is arrived at. Table 2.5 gives the nomenclature and Fig. 2.5 the listing and results.

Example

450,000 lb/hr of saturated steam at 1000 psia flows in a pipe of inner diameter 9.562 in. Determine the orifice size to limit the differential head to 500 in. WC.

TABLE 2.5 Nomenclature for Program 2.5

Nomenclature	Program symbol	Description and units
β	Y	Ratio d/d_i
K	Z	Factor defined in equation (1)
d_i	D	Pipe inner diameter (in.)
d	I	Orifice diameter (in.)
h	H	Differential head (in. WC)
M	M	Factor defined in equation (1)
P	P	Steam pressure (psia)
Re	RE	Reynolds number
t	T	Steam temperature (°F)
μ	—	Viscosity (lb/ft hr)
v	V	Specific volume (ft^3/lb)
W	W	Steam flow (lb/hr)

```
10 PRINT"ORIFICE FOR STEAM FLOW"
20 INPUT"STEAM FLOW LB/H,PRESSURE-PSIA,TEMP(IF SATURATED INPUT 0 AND PROGRAM COM
PUTES THE VALUE)=";WS,P,T
30 INPUT"TUBE ID-IN,DIFFERENTIAL HEAD IN WC=";D,H
40 IF T=0 THEN T=115*P^.225
50 GOSUB 500
60 M1=WS/(2837*D*D*SQR(.016*H/V))
70 RE=15.2*WS/(D*(.016+.000058*T))
80 GOSUB 600
90 I=D*Y
95 PRINT"RESULTS"
100 PRINT" "
110 PRINT"ORIFICE SIZING"
120 PRINT" "
130 PRINT"STEAM FLOW-LB/H=";WS;"TEMP=";T;"PRESS-PSIA=";P
140 PRINT" "
150 PRINT"DIFF HEAD-IN WC=";H;"PIPE ID-IN=";D;"ORIFICE-IN=";I
160 PRINT" "
170 PRINT"STEAM SP VOL-CU FT/LB=";V
175 PRINT" "
180 END
500 T=273.1+(T-32)/1.8:P=P/14.696
510 K=80870!/T/T:L=10^K*(-2641.62/T):M=1.89+L:N=M*P*P/T/T
520 O=2+(372420!/T/T):Q=O*L:R=1.89+Q:U=(.21828*T-126970!/T):V=2*U*R-(M/T)*126460
!
530 W=82.54-162460!/T:Y=2*W*R-(M/T)*162460!
540 V=(((U*M*N+W)*N/P+1)*M+4.5504*T/P)*.0160185
550 P=14.696*P:T=(T-273.15)*1.8+32
560 RETURN
590 P=14.696*P:T=(T-273.15)*1.8+32
600 X=.5:Y=.5
610 Z=.593+.4*Y^4+(.0015*Y^.5+.012*Y^4)*1000/RE^.5
620 F=Z*Y*Y
630 IF ABS((M1-F)/M1)<.03 THEN 650
640 X=.5*X:Y=Y+SGN(M1-F)*X:GOTO 610
650 RETURN

ORIFICE FOR STEAM FLOW
STEAM FLOW LB/H,PRESSURE-PSIA,TEMP(IF SATURATED INPUT 0 AND PROGRAM COMPUTES THE
 VALUE)=? 450000,1000,0
TUBE ID-IN,DIFFERENTIAL HEAD IN WC=? 9.562,500
RESULTS

ORIFICE SIZING

STEAM FLOW-LB/H= 450000 TEMP= 544.0339 PRESS-PSIA= 999.9999

DIFF HEAD-IN WC= 500 PIPE ID-IN= 9.562 ORIFICE-IN= 7.1715

STEAM SP VOL-CU FT/LB= .4449299
```

FIG. 2.5 Listing of program 2.5, with results.

Solution

Key in the program. In the RUN mode, the following data are input in the same order given here.

Steam flow (450,000)
Pressure, psia (1000)
Temperature (0) (if superheated, input actual value)
Tube inner diameter (9.562)
Differential head (500)

Once these are fed in, the computer goes on to solve for the orifice size and prints out the results.

It is seen from Fig. 2.5 that steam temperature is 544°F, volume is 0.445 ft^3/lb, and orifice size is 7.17 in.

PROGRAM 2.6 SIZING ORIFICES FOR WATER FLOW

Input

Flow (lb/hr or gpm)
Temperature
Pipe inner diameter
Differential head

Output

Flow (lb/hr)
Flow (gpm)
Orifice diameter

The program sizes orifices for metering water flow in pipes. The effect of temperature is considered, as it is important in boiler applications. Flow can be input either in lb/hr or in gpm, and the program is designed to compute both the values. A quick convergence routine is adopted to solve for the orifice size.

Theory

The basic equation for sizing orifices for liquid flow* is

$$q = \frac{19.636Cd^2(h/12)^{0.5}}{(1-\beta^4)^{0.5}} \tag{1}$$

*Ingersoll Rand, *Cameron Hydraulic Data*, 16th edition, Woodcliff Lake, New Jersey, pp. 2-8

where $\beta = d/d_i$. Rewriting, we have

$$M = \frac{\beta^2}{\sqrt{1-\beta^4}} = \frac{q}{19.636Cd_i^2(h/12)^{0.5}} \quad (2)$$

Also

$$q = 0.125Wv \quad (3)$$

The specific volume v is estimated from the equations

$$v = 0.016 \qquad \text{for } t \leqslant 120°F \quad (4a)$$

$$v = 0.0149 \exp(0.000555t) \qquad \text{for } t > 120°F \quad (4b)$$

Table 2.6 gives the nomenclature. The program computes M from equation (2) and arrives at β through a trial-and-error procedure. The discharge coefficient C is 0.61. Two examples illustrate the use of the program, a listing of which appears in Fig. 2.6.

Example 1

A total of 50,000 lb/hr of water at 108°F flows in a boiler plant line of size 3 in. sch 80. Determine the flow in gpm and the orifice size to limit the differential head to 100 in. WC.

TABLE 2.6 Nomenclature for Program 2.6

Nomenclature	Program symbol	Description and units
β	B	Ratio of orifice to pipe diameter
C	—	Discharge coefficient
d_i	D	Pipe inner diameter (in.)
d	I	Orifice diameter (in.)
h	H	Differential head (in. WC)
M	M	Factor defined in equation (2)
q	Q	Flow (gpm)
t	T	Water temperature (°F)
v	V	Specific volume (ft^3/lb)
W	W	Flow (lb/hr)

```
10 PRINT"ORIFICE SIZING FOR WATER FLOW"
15 PRINT"FLOW MAY BE GIVEN EITHER IN LB/H OR IN GPM;PROGRAM COMPUTES BOTH;INPUT
Ø FOR THE UNKNOWN VALUE"
2Ø INPUT"FLOW -LB/H,FLOW-GPM,WATER TEMP,PIPE ID-IN,DIFF HEAD-IN WC=";W,Q,T,D,H
3Ø IF T<=12Ø THEN V=.Ø16:GOTO 5Ø
4Ø V=.Ø149*EXP(.ØØØ555*T)
5Ø IF W=Ø THEN M=Q/(19.636*.61*D*D*(H/12)^.5)
6Ø IF Q=Ø THEN M=.125*W*V/(19.636*.61*D*D*(H/12)^.5)
7Ø GOSUB 5ØØ
8Ø I?B*D
9Ø IF W=Ø THEN W=8*Q/V
1ØØ IF Q=Ø THEN Q=.125*W*V
11Ø PRINT" "
12Ø PRINT"ORIFICE SIZING FOR WATER"
13Ø PRINT" "
14Ø PRINT"FLOW-LB/H=";W;"FLOW-GPM=";Q;"TEMP=";T
15Ø PRINT" "
16Ø PRINT"PIPE ID-IN=";D;"ORIFICE DIA-IN=";I;"DIFF HEAD-IN WC=";H
165 PRINT" "
17Ø END
5ØØ L=Ø:Z=1:R=.5
51Ø Z=.5*(L+Z)
52Ø F=Z*Z/(1-Z^4)^.5
53Ø IF ABS(M-F)<.Ø1 THEN 55Ø
54Ø R=.5*R:Z=Z+SGN(M-F)*R:GOTO 52Ø
55Ø B=Z:RETURN
```

```
FLOW -LB/H,FLOW-GPM,WATER TEMP,PIPE ID-IN,DIFF HEAD-IN WC=? 5ØØØØ,Ø,1Ø8,2.9,1ØØ

ORIFICE SIZING FOR WATER
                                    " "
FLOW-LB/H= 5ØØØØ FLOW-GPM= 1ØØ TEMP= 1Ø8

PIPE ID-IN= 2.9 ORIFICE DIA-IN= 1.6539Ø6 DIFF HEAD-IN WC= 1ØØ


ORIFICE SIZING FOR WATER FLOW
FLOW MAY BE GIVEN EITHER IN LB/H OR IN GPM;PROGRAM COMPUTES BOTH;INPUT Ø FOR THE
 UNKNOWN VALUE
FLOW -LB/H,FLOW-GPM,WATER TEMP,PIPE ID-IN,DIFF HEAD-IN WC=? Ø,13,21Ø,1.Ø4,1ØØ

ORIFICE SIZING FOR WATER

FLOW-LB/H= 6211.98 FLOW-GPM= 13 TEMP= 21Ø

PIPE ID-IN= 1.Ø4 ORIFICE DIA-IN= .6Ø125 DIFF HEAD-IN WC= 1ØØ
```

FIG. 2.6 Listing of program 2.6, with results.

Solution

Key in the program. In the RUN mode, the screen asks for the following data in the same order given here.

Flow, lb/hr (50,000)
Flow, gpm (0) (if unknown, input 0)
Pipe inner diameter (2.9)
Differential head (100)
Water temperature (108)

Once these are fed in, the computer solves through a trial-and-error procedure and prints out the results. It is seen from Fig. 2.6 that flow (gpm) is 100 and the orifice is 1.65 in.

Example 2

A total of 13 gpm of water at 210° F flows in a pipe of inner diameter 1.05 in. Determine the flow (lb/hr) and orifice size to limit the differential head to 100 in. WC.

Solution

In the RUN mode the display asks for the same set of data as before. Input 0 for flow (lb/hr). The computer solves for flow as 6307 lb/hr, and orifice size is 0.6 in.

PROGRAM 2.7 PRESSURE DROP OF AIR AND FLUE GAS OVER FINNED TUBES

Input

Gas flow
Temperature
Tube OD
Length
Number wide deep
Fin density, height, and thickness
Transverse and longitudinal pitch

Output

Mass velocity of gas
Linear velocity
Gas pressure drop (in. WC)

Remarks

The correlation of Briggs and Young is used. It is valid for a staggered arrangement and solid fins and gives conservative estimates.

The program computes the pressure drop of air and flue gases flowing over circumferentially finned tubes arranged in a staggered fashion as obtained in economizers and waste heat boilers. The correlation used is that of Briggs and Young, which gives a conservative estimate. Results include the gas pressure drop, gas mass, and linear velocities.

Theory

The equation for gas pressure drop* is

$$\Delta P_g = 18.93 Re^{-0.316} \left(\frac{S_T}{d}\right)^{-0.927} \left(\frac{S_T}{S_L}\right)^{0.515} \frac{G^2 N_h}{g_c \rho_g} \tag{1}$$

The equation may be simplifed after substituting for the gas molecular weight as

$$\Delta P_g = \frac{54 \times 10^{-11} G^{1.684} d^{0.611} (136 + 0.5283t) N_h}{S_T^{0.412} S_L^{0.515}} \tag{2}$$

$$G = \frac{W}{N_w L[(S_T/12) - A_0]} \tag{3}$$

$$A_0 = \frac{d}{12} + \frac{nbh}{6} \tag{4}$$

Linear velocity

$$V = \frac{G}{3600\rho_g} = \frac{G(460 + t)}{144{,}000} \tag{5}$$

using the equation $\rho_g = 40/(460 + t)$ for the density of common flue gases and air. Table 2.7 gives the nomenclature and Fig. 2.7 the listing of the program.

---

*V. Ganapathy, *Applied Heat Transfer*, Pennwell Books, Tulsa, Oklahoma, 1982, p. 515.

TABLE 2.7 Nomenclature for Program 2.7

Nomenclature	Program symbol	Description and units
A_0	A	Obstruction area (ft^2/ft)
b	B	Fin thickness (in.)
d	D	Tube outer diameter (in.)
G	G	Gas mass velocity ($lb/ft^2/hr$)
h	H	Fin height (in.)
L	L	Finned tube length (ft)
MW	—	Gas molecular weight
n	N	Fins per in.
N_h	K	Number of rows deep
N_w	M	Number of tubes wide
ΔP_g	P	Gas pressure drop (in. WC)
S_T	X	Transverse pitch (in.)
S_L	Y	Longitudinal pitch (in.)
t	T	Average air and gas temperature (°F)
V	V	Gas velocity (ft/sec)
W	W	Gas flow (lb/hr)
μ	—	Gas viscosity (lb/ft hr)

Example

In a waste heat boiler, the following data are noted:

1. Gas flow = 75,000 lb/hr
2. Gas temperature (average) = 1000°F
3. Transverse pitch = 4.0 in.
4. Longitudinal pitch = 3.6 in.
5. Tube outer diameter = 2 in.
6. Length = 9 ft
7. Number of tubes wide = 16
8. Number of tubes deep = 10
9. Fins per in. = 3
10. Fin height = 0.75 in.
11. Fin thickness = 0.059 in.

Determine the gas pressure drop and the linear and mass velocities.

```
10 PRINT"GAS PR DROP OVER FINNED TUBES"
20 PRINT" SOLID FINS-STAGGERED ARRANGEMENT"
30 INPUT"GAS FLOW,GAS TEMP,TUBE OD,TR PITCH,LONG PITCH,TUBE LENGTH=";W,T,D,X,Y,L

40 INPUT"FINS/IN,FIN HT,FIN THK,NO OF TUBES WIDE,DEEP=";N,H,B,M,K
50 A=(D/12)+N*B*H/6:G=W/(M*L*((X/12)-A)):V=G*(460+T)/144000!
60 P=54*10^(-11)*G^1.684*D^.611*(136+.5283*T)*K/(X^.412*Y^.515)
70 PRINT" "
80 PRINT"GAS FLOW=";W;"GAS TEMP=";T
90 PRINT" "
100 PRINT"MASS VEL=";G;"LIN VEL=";V
110 PRINT" "
120 PRINT"TUBE OD=";D;"TR PITCH=";X;"LONG PITCH=";Y
130 PRINT" "
140 PRINT"NO WIDE=";M;"NO DEEP=";K;"FINS/IN=";N;"FIN HT=";H;"THK=";B
150 PRINT" "
160 PRINT"GAS PR DROP-IN WC=";P
170 END

GAS PR DROP OVER FINNED TUBES
 SOLID FINS-STAGGERED ARRANGEMENT
GAS FLOW,GAS TEMP,TUBE OD,TR PITCH,LONG PITCH,TUBE LENGTH=? 75000,1000,2,4,3.6,9

FINS/IN,FIN HT,FIN THK,NO OF TUBES WIDE,DEEP=? 3,.75,.059,16,10

GAS FLOW= 75000 GAS TEMP= 1000

MASS VEL= 3603.344 LIN VEL= 36.5339

TUBE OD= 2 TR PITCH= 4 LONG PITCH= 3.6

NO WIDE= 16 NO DEEP= 10 FINS/IN= 3 FIN HT= .75 THK= 5.900001E-02

GAS PR DROP-IN WC= 1.561834
```

FIG. 2.7 Listing of program 2.7, with results.

Solution

Key in the program. In the RUN mode, the screen asks for the data in the same order given here; they are fed in, then the results are printed out. It is seen that the gas mass velocity is 3603 lb/ft^2 hr and the linear velocity is 36 ft/sec; the gas pressure drop is 1.56 in. WC.

PROGRAM 2.8 PRESSURE DROP OF AIR AND FLUE GAS OVER BARE TUBES

Input

Arrangement (inline or staggered)
Tube OD
Transverse and longitudinal pitch
Number of tubes wide and deep
Tube length
Gas flow and temperature

Output

Mass velocity
Linear velocity
Pressure drop in (in. WC)

Remarks

The program is valid for inline and staggered tube bundles.

The program calculates the pressure drop of air and flue gas flowing over a bare tube bundle arranged either inline or in a staggered fashion. The inputs are the coil configuration, gas flow, and temperature. The output includes the mass and linear velocity and the pressure drop.

Theory

The following equations are used to arrive at the gas pressure drop*:

$$\Delta P_g = \frac{9.3 \times 10^{-10} f G^2 N_H}{\rho_g} \tag{1}$$

where, for inline arrangement for $S_T/d = 1.5$ to 4.0 and for $2000 < Re < 40{,}000$,

$$f = Re^{-0.15} \left\{ 0.044 + \frac{0.08 S_L/d}{[(S_T/d) - 1]^{(0.43 + 1.13d/S_L)}} \right\} \tag{2}$$

*V. Ganapathy, *Applied Heat Transfer*, Pennwell Books, Tulsa, Oklahoma, 1982, p. 542.

TABLE 2.8 Nomenclature for Program 2.8

Nomenclature	Program symbol	Description and units
d	D	Tube outer diameter (in.)
f	F	Friction factor
G	G	Gas mass velocity, $Lb/ft^2 h$
L	H	Tube length (ft)
N_w	N	Number of tubes wide
N_H	K	Number of tubes deep
ΔP_g	P	Gas pressure drop (in. WC)
Re	R	Reynolds number
S_T	S	Transverse pitch (in.)
S_L	L	Longitudinal pitch (in.)
t	T	Gas temperature (°F)
μ	V	Gas viscosity (lb/ft hr)
V	C	Gas velocity (ft/sec)
W	W	Gas flow (lb/hr)
—	A	Arrangement factor: 1, for inline, 2 for staggered

For a staggered arrangement for $S_T/d = 1.5$ to 4.0,

$$f = Re^{-0.16} \left\{ 0.25 + \frac{0.1175}{[(S_T/d) - 1]^{1.08}} \right\} \tag{3}$$

where

$$\text{Reynolds number } Re = \frac{Gd}{12\mu} \tag{4}$$

$$G = \frac{12W}{N_w L(S_T - d)} \tag{5}$$

For air and common flue gases, viscosity may be estimated as

$$\mu = 0.0513 + 3.66 \times 10^{-5} t \qquad \text{for } t > 800°F \tag{6a}$$

$$\mu = 0.0405 + 5.4 \times 10^{-5} t \qquad \text{for } t \leqslant 800°F \tag{6b}$$

```
10 PRINT"GAS PR DROP OVER BARE TUBE BUNDLES"
20 PRINT"INLINE OR STAGGERED ARRANGEMENT"
30 INPUT"ARRANGEMENT FACTOR-1 FOR INLINE-2 FOR STAGGERED=";A
40 INPUT"TUBE OD,TR PITCH,LONG PITCH-IN=";D,S,L
50 INPUT"NO OF TUBES WIDE,NO DEEP,LENGTH -FT=";N,K,H
60 INPUT"GAS FLOW,GAS TEMP=";W,T
70 IF T<=800 THEN V=.0405+.000054*T:GOTO 90
80 V=.0513+.0000366*T
90 G=12*W/(N*(S-D)*H):R=G*D/12/V:C=G*(460+T)/144000!
100 IF A=2 THEN F=R^(-.16)*(.25+.1175/((S/D)-1)^1.08):GOTO 120
110 F=R^(-.15)*(.044+.08*(L/D)/((S/D)-1)^(.43+1.13*D/L))
120 P=.2325*10^(-10)*F*G*G*K*(460+T)
130 PRINT" "
140 PRINT"GAS PR DROP CALCULATIONS"
150 PRINT" "
160 PRINT"GAS FLOW=";W;"GAS TEMP=";T
170 PRINT" "
180 PRINT"GAS MASS VEL-LB/FT2H=";G;"LIN VEL-FT/S=";C
190 PRINT" "
200 PRINT"TUBE OD=";D;"TR PITCH=";S;"LONG PITCH=";L;"TUBE LENGTH=";H
210 PRINT" "
220 PRINT"TUBES WIDE=";N;"NO DEEP=";K;"GAS PR DROP-IN WC=";P
230 PRINT" "
240 IF A=1 THEN PRINT "INLINE COIL":GOTO 260
250 PRINT"STAGGERED COIL"
260 END

GAS PR DROP OVER BARE TUBE BUNDLES
INLINE OR STAGGERED ARRANGEMENT
ARRANGEMENT FACTOR-1 FOR INLINE-2 FOR STAGGERED=? 1
TUBE OD,TR PITCH,LONG PITCH-IN=? 2,3.5,4
NO OF TUBES WIDE,NO DEEP,LENGTH -FT=? 18,30,7.1
GAS FLOW,GAS TEMP=? 150000,1000

GAS PR DROP CALCULATIONS

GAS FLOW= 150000 GAS TEMP= 1000

GAS MASS VEL-LB/FT2H= 9389.672 LIN VEL-FT/S= 95.20084

TUBE OD= 2 TR PITCH= 3.5 LONG PITCH= 4 TUBE LENGTH= 7.1

TUBES WIDE= 18 NO DEEP= 30 GAS PR DROP-IN WC= 5.316179

INLINE COIL
```

FIG. 2.8 Listing of program 2.8, with results.

The density ρ_g may be estimated as

$$\rho_g = \frac{40}{460 + t} \tag{7}$$

and velocity as

$$V = \frac{G}{3600\rho_g} \tag{8}$$

A listing of the program appears in Fig. 2.8, and Table 2.8 gives the nomenclature.

Example

Determine the pressure drop of flue gases flowing over a staggered tube bundle with the following data:

1. Arrangement = 1 (input 1 for inline and 2 for staggered)
2. Tube outer diameter = 2 in.
3. Transverse pitch = 4.0 in.
4. Longitudinal pitch = 3.2 in.
5. Number of tubes wide = 18
6. Number deep = 30
7. Tube length = 7.1 ft
8. Gas flow = 150,000 lb/hr
9. Gas temperature = 1000°F

Solution

In the RUN mode, the screen asks for the data in the same order given here; they are fed in. It is seen from Fig. 2.8 that the gas mass velocity is 9389 lb/ft^2 hr, velocity is 96 ft/sec, and gas pressure drop is 5.31 in. WC.

3

Heat Transfer Calculations

PROGRAM 3.1 HEAT TRANSFER COEFFICIENT FOR STEAM INSIDE TUBES

Input

Steam pressure
Temperature
Tube inner diameter
Flow

Output

Steam temperature
Heat transfer coefficient

Remarks

For saturated steam, if the temperature is not known, the input of 0 will enable the computer to calculate the saturation temperature and then the heat transfer coefficient.

With this program, one can obtain the steam heat transfer coefficient inside tubes given only the steam flow, pressure, temperature, and tube inner diameter. Thermal and transport properties have been evaluated, and correlations have been developed to consider their effect. Hence, there is no need to refer to such properties as viscosity, thermal conductivity, and specific heat. The program handles saturated and superheated steam.

Theory

The heat transfer coefficient for flow of fluids inside tubes in the turbulent regime, which is the case with most of the situations in steam plants, is obtained from the familiar Dittus-Boelter correlation

$$NU = 0.023Re^{0.8}Pr^{0.4} \tag{1}$$

where the Nusselt number

$$NU = \frac{h_i d_i}{12k} \tag{2}$$

Prandtl number is

$$Pr = \frac{\mu c_p}{k} \tag{3}$$

and the Reynolds number

$$Re = \frac{15.2w}{d_i \mu} \tag{4}$$

The properties c_p, μ, and k are estimated at the fluid bulk temperature T.

Equation (1) may be rewritten as follows after substituting equations (2) through (4):

$$h_i = 2.44 \frac{w^{0.8}}{d_i^{1.8}} F \tag{5}$$

where

$$F = \frac{k^{0.6} c_p^{0.4}}{\mu^{0.4}} \tag{6}$$

The factor F may be estimated for saturated steam as

$$F = 0.172 + 0.079X - 0.007297X^2 + 0.000257X^3 \tag{7}$$

where X = 0.01P, and, for superheated steam when $P < 200$ psia

$$F = 0.244 + 0.000132(T - 327) \tag{8}$$

When $P > 200$ psia

$$F = 0.343 + 0.071M - 0.043N - 0.058MN + 0.028M^2 + 0.041N^2 \tag{9}$$

where M = 3.32 log P − 10 and N = 0.005T − 4.

The program also computes the saturation temperature T if zero is input for steam temperature using the equation

$$T = 115P^{0.225} \tag{10}$$

A listing of the program appears in Fig. 3.1 and Table 3.1 gives the nomenclature.

Example 1

Calculate the tubeside heat transfer coefficient in a superheater when 6500 lb/hr of saturated steam at 1600 psia flows through a tube with inner diameter 1.706 in.

```
5 PRINT"STEAM SIDE HEAT TRANSFER COEFFICIENT"
10 INPUT"STM PRESSURE,PSIA,STEAM TEMP(if saturated input zero)";P,T
15 S=115*P^.225
20 INPUT"TUBE ID-in,FLOW/TUBE-lb/h=";I,W
30 IF T=0 GOTO 70
35 IF P<200 THEN 80
40 X=1.434*LOG(P)-10:Y=.005*T-4
50 F=.343+.071*X-.043*Y-.058*X*Y+.028*X*X+.041*Y*Y
60 GOTO 90
70 X=.01*P:F=.172+.079*X-.007297*X*X+.000257*X^3:S=115*P^.225:T=S
75 GOTO 90
80 F=.244+.000132*(T-327)
90 H=2.44*W^.8*F/I^1.8
95 PRINT" "
100 PRINT"RESULTS"
105 PRINT" "
110 PRINT"STEAM PRESSURE-PSIA=";P;"STEAM TEMPERATURE-F=";T
115 PRINT" "
120 PRINT"FLOW-LB/H=";W;"TUBE ID-IN =";I
125 PRINT" "
130 PRINT"STEAM HEAT TRANSFER COEFFT-BTU/FT2HF=";H
135 PRINT" "
140 END

RESULTS

STEAM PRESSURE-PSIA= 1600 STEAM TEMPERATURE-F= 604.8181

FLOW-LB/H= 6500 TUBE ID-IN = 1.706

STEAM HEAT TRANSFER COEFFT-BTU/FT2HF= 650.1266


STEAM SIDE HEAT TRANSFER COEFFICIENT
STM PRESSURE,PSIA,STEAM TEMP(if saturated input zero)? 900,650
TUBE ID-in,FLOW/TUBE-lb/h=? 1.706,6500

RESULTS

STEAM PRESSURE-PSIA= 900 STEAM TEMPERATURE-F= 650

FLOW-LB/H= 6500 TUBE ID-IN = 1.706

STEAM HEAT TRANSFER COEFFT-BTU/FT2HF= 389.5728
```

FIG. 3.1 Listing of program 3.1, with results.

Solution

Key in the program. In the RUN mode, the screen asks for the steam pressure (1600), steam temperature (0, as saturated), tube

TABLE 3.1 Nomenclature for Program 3.1

Nomenclature	Program symbol	Description and units
c_p	—	Specific heat (Btu/lb °F)
d_i	I	Tube inner diameter (in.)
F	F	Factor defined in equation (6)
k	—	Thermal conductivity (Btu/fthr °F)
NU	—	Nusselt number
P	P	Steam pressure (psia)
Pr	—	Prandtl number
Re	—	Reynolds number
T	T	Steam temperature (°F)
w	W	Steam flow (lb/hr)
μ	—	Steam viscosity (lb/ft hr)

inner diameter (1.706), and flow (6500), which are input. The computer goes on to calculate the saturation temperature and the heat transfer coefficient and prints out the results. It is seen that saturation temperature is 604°F and heat transfer coefficient is 650 Btu/ft^2 hr °F.

Example 2

What is the heat transfer coefficient when 6500 lb/hr of superheated steam at 900 psia and 650°F flows through the same tube?

Solution

In the RUN mode, the appropriate data are fed in. Temperature is now 650. It is seen that the heat transfer coefficient is 390 Btu/ft^2 hr °F.

Reference

V. Ganapathy, *Applied Heat Transfer*, Pennwell Books, Tulsa, Oklahoma, 1982, p. 433.

PROGRAM 3.2 HEAT TRANSFER COEFFICIENT FOR WATER INSIDE TUBES

Input

Water flow (lb/hr)
Temperature
Tube inner diameter

Output

Velocity
Heat transfer coefficient

Remarks

Effect of pressure is not significant for water, and hence the program may be used for high-pressure water.

This program computes the heat transfer coefficient for water flowing inside tubes. The input is merely the tube inner diameter, water flow, and temperature. Thermal and transport properties are taken into account, so reference to water properties is not required.

Theory

The same theory discussed in the program 3.1 is applicable. Factor F may be written as

$$F = 10^{(-1.318 + .493 \log t)} \tag{1}$$

Then,

$$h_i = \frac{2.44 w^{0.8} F}{d_i^{1.8}} \tag{2}$$

Velocity

$$V = \frac{0.05 W v}{d_i^{2}} \tag{3}$$

Specific volume v is computed using correlations discussed in program 2.2.

Example

In a boiler economizer, 10,000 lb/hr of water at 400°F flows inside a tube of inner diameter 1.7 in. What are the tubeside heat transfer coefficient and water velocity?

Solution

Key in the program, a listing of which appears in Fig. 3.2; the nomenclature is listed in Table 3.2. In the RUN mode, the display asks for the flow, temperature, and tube inner diameter, which are fed in. The printout shows a heat transfer coefficient of 1370 Btu/ft^2 hr °F and a water velocity of 3.218 ft/sec.

```
10 PRINT "HEAT TRANSFER COEFFICIENT FOR WATER IN TUBES"
20 INPUT"WATER FLOW IN LB/H PER TUBE,TUBE ID ,IN AND WATER TEMP";W,D,T
30 H=2.44*W^.8*(10^(.214*LOG(T)-1.318))/D^1.8
40 IF T <=120 THEN V=.016:GOTO 60
50 V=.0149*EXP(.000555*T)
60 S=.05*W*V/D/D
70 PRINT" "
80 PRINT"RESULTS"
85 PRINT" "
90 PRINT"WATER FLOW-LB/H=";W;"WATER TEMP-F=";T
95 PRINT" "
100 PRINT"WATER VELOCITY-FT/S=";S;"HEAT TRANSFER COEF-BTU/FT2HF=";H
105 PRINT" "
110 PRINT"TUBE ID-IN=";D
115 PRINT" "
120 END

HEAT TRANSFER COEFFICIENT FOR WATER IN TUBES
WATER FLOW IN LB/H PER TUBE,TUBE ID ,IN AND WATER TEMP? 10000,1.7,400

RESULTS

WATER FLOW-LB/H= 10000 WATER TEMP-F= 400

WATER VELOCITY-FT/S= 3.218636 HEAT TRANSFER COEF-BTU/FT2HF= 1370.109

TUBE ID-IN= 1.7
```

FIG. 3.2 Listing of program 3.2, with results.

TABLE 3.2 Nomenclature for Program 3.2

Nomenclature	Program symbol	Description and units
d_i	I	Tube ID (in.)
F	F	Factor discussed in equation (1)
h_i	H	Heat transfer coefficient (Btu/ft^2 hr °F)
t	T	Water temperature (°F)
V	S	Water velocity (ft/sec)
v	V	Specific volume (ft^3/lb)
w	W	Water flow (lb/hr)

PROGRAM 3.3 HEAT TRANSFER COEFFICIENT FOR AIR AND FLUE GAS INSIDE TUBES

Input

Flow (lb/hr)
Gas/air temperature
Tube inner diameter

Output

Density
Volume (acfm)
Velocity
Heat transfer coefficient

Remarks

Air or flue gas is at low pressure. The program gives good estimates for common flue gases (products of combustion of fossil fuels).

The program computes the heat transfer coefficient for air and common flue gases flowing inside tubes. Input are the flow, pressure, temperature, and tube inner diameter (see Table 3.3).

Theory

The theory discussed in program 3.1 is applicable. Factor F may be written as

TABLE 3.3 Nomenclature for Program 3.3

Nomenclature	Program symbol	Description and units
d_i	I	Tube inner diameter (in.)
F	F	Factor defined in equation (1)
h_i	H	Heat transfer coefficient ($Btu/ft^2\ hr\ ^\circ F$)
P	P	Pressure (psia)
t	T	Fluid temperature ($^\circ$F)
V	V	Velocity (ft/sec)
w	W	flow (lb/hr)
ρ	D	Density (lb/ft^3)

$$F = 0.163 + 0.000042(t - 200) \tag{1}$$

F does not vary significantly with pressure up to 300 psig.* Then,

$$h_i = \frac{2.44w^{0.8}F}{d_i^{1.8}} \tag{2}$$

The tubeside velocity V is computed using a molecular weight of 29 for air/flue gas in the evaluation of density.

$$\rho = \frac{2.7P}{460 + t} \tag{3}$$

$$V = \frac{0.05W}{\rho d_i^2} \tag{4}$$

Example

Determine the heat transfer coefficient when 200 lb/hr of air at 15 psia and 700°F flows inside a tube of inner diameter 1.7 in.

Solution

Key in the program. In the RUN mode, the screen asks for the data, which are fed in. It is seen from Fig. 3.3 that the heat transfer coefficient is 11.97 $Btu/ft^2\ hr\ ^\circ F$ and the velocity is 99 ft/sec.

*V. Ganapathy, *Applied Heat Transfer*, Pennwell Books, Tulsa, Oklahoma, 1982, pp. 441-443.

```
1Ø PRINT"HEAT TRANSFER COEFFICENT FOR AIR/FLUE GAS IN TUBES"
2Ø INPUT"GAS FLOW-LB/H,TEMPERATURE,PRESSURE-PSIA,TUBE ID-IN=";W,T,P,I
3Ø D=2.7*P/(46Ø+T):Q=W/6Ø/D:V=.Ø5*W/I/I/D
4Ø H=2.44*W^.8*(.163+.ØØØØ42*(T-2ØØ))/I^1.8
5Ø PRINT" "
6Ø PRINT"GAS HEAT TRANSFER COEFFICIENT"
7Ø PRINT" "
8Ø PRINT"GAS FLOW-LB/H=";W;"FLOW-ACFM=";Q;"TEMP-F=";T;"PRESS-PSIA=";P
9Ø PRINT" "
1ØØ PRINT"TUBE ID-IN=";I;"VELOCITY-FT/S=";V;"DENSITY-LB/CU FT=";D
11Ø PRINT" "
12Ø PRINT"GAS HT TR COEFF-BTU/FT2HF=";H
13Ø PRINT" "
14Ø END

HEAT TRANSFER COEFFICENT FOR AIR/FLUE GAS IN TUBES
GAS FLOW-LB/H,TEMPERATURE,PRESSURE-PSIA,TUBE ID-IN=? 2ØØ,7ØØ,15,1.7

GAS HEAT TRANSFER COEFFICIENT

GAS FLOW-LB/H= 2ØØ FLOW-ACFM= 95.47325 TEMP-F= 7ØØ PRESS-PSIA= 15

TUBE ID-IN= 1.7 VELOCITY-FT/S= 99.1Ø718 DENSITY-LB/CU FT= 3.491379E-Ø2

GAS HT TR COEFF-BTU/FT2HF= 11.97357
```

FIG. 3.3 Listing of program 3.3, with results.

PROGRAM 3.4 HEAT TRANSFER COEFFICIENT WITH FINNED TUBE BUNDLES

Input

Gas or air flow
Temperature
Tube OD
Tube length
Number of tubes wide
Transverse pitch
Fin density
Fin height
Fin thickness

Output

Gas mass velocity
Linear velocity
Fin area (ft^2/ft)

Total Surface area of tube (ft^2/ft)
Gas heat transfer coefficient
Fin effectiveness

Remarks

This technique, which uses the equation of Briggs and Young, is for solid finned tubes in a staggered arrangement. It gives conservative estimates.

The program calculates the heat transfer coefficient in circumferentially finned tubes arranged in staggered fashion, when air or flue gases flow over them. Input are gas flow, temperature, and fin configuration, and the results include fin area, total area, gas heat transfer coefficient, and fin efficiency and effectiveness.

Theory

The following equations are used to arrive at the gas heat transfer coefficient and other related parameters:

$$\frac{h_g d}{12k} = 0.134 \left(\frac{Gd}{12\mu}\right)^{0.681} \left(\frac{\mu C_p}{k}\right)^{0.33} \left(\frac{S}{h}\right)^{0.2} \left(\frac{S}{b}\right)^{0.113} \tag{1}$$

Simplifying, we have

$$h_g = 0.295 \frac{G^{0.681}}{d^{0.319}} \left(\frac{k^{0.67} C_p{}^{0.33}}{\mu^{0.351}}\right)\left(\frac{S^{0.313}}{h^{0.2} b^{0.113}}\right) \tag{2}$$

The factor $F = k^{0.67} c_p{}^{0.33}/\mu^{0.351}$ has been evaluated for common flue gases, which is also reasonably accurate for air, and may be written as

$$F = 0.125 + 0.00004(t - 400) \tag{3}$$

$$\text{Gas mass velocity } G = \frac{W_g}{N_w\,[(s_t/12) - A_0\,]} \tag{4}$$

$$A_0 = \frac{d}{12} + \frac{nbh}{6} \tag{5}$$

$$\text{Fin spacing } S = \frac{1}{n} - b \tag{6}$$

$$\text{Fin efficiency } \phi = \frac{1}{1 + m^2h^2\,[(d + 2h)/d]^{0.5}/432} \tag{7}$$

$$\text{Fin effectiveness} = 1 - \frac{(1-\phi)A_f}{A_t} = \eta \tag{8}$$

$$m = \left(\frac{K_m b}{24h_g}\right)^{0.5} \tag{9}$$

$$\text{Fin area } A_f = \frac{\pi n}{24}(4dh + 4h^2 + 2bd + 4bh) \tag{10}$$

$$\text{Total area of finned tube } A_t = A_f + \frac{\pi d}{12}(1 - nb) \tag{11}$$

Once h_g is obtained, the corrected outside heat transfer coefficient is given by ηh_g. Table 3.4 gives the nomenclature; Fig. 3.4 gives the listing and results.

Example

In a finned tube bundle, the following data are noted:

1. Gas flow - 150,000 lb/hr
2. Average gas temperature = 525°F
3. Tube OD = 3.5 in.
4. Tube length = 10.5 ft
5. Number of tubes wide = 12
6. Transverse pitch = 6 in.
7. Fins per inch = 3
8. Fin height = 0.75 in.
9. Fin thickness = 0.059 in.

Evaluate the gas heat transfer coefficient and other related parameters.

Solution

Key in the program. In the RUN mode, the screen asks for the data in the same order given here, and they are fed in. The results are shown in Fig. 3A. It is seen that the gas mass velocity is 6393 lb/ft^2 hr, linear velocity is 43 ft/sec, gas heat transfer coefficient is

TABLE 3.4 Nomenclature for Program 3.4

Nomenclature	Program symbol	Description and units
A_f	F	Fin area (ft^2/ft)
A_t	S	Total external area (ft^2/ft)
b	B	Fin thickness (in.)
c_p		Gas specific heat (Btu/lb °F)
d	D	Tube OD (in.)
G	G	Gas mass velocity ($lb/ft^2/hr$)
h	H	Fin height (in.)
h_g	O	Gas heat transfer coefficient (Btu/ft^2 hr °F)
k	—	Gas thermal conductivity (Btu/ft hr °F)
L	L	Tube length (ft)
n	N	Fin density (fins per inch)
N_w	M	Number of tubes wide
s		Fin spacing (in.)
s_t	X	Transverse pitch (in.)
t	T	Gas temperature (°F)
V	V	Gas velocity (ft/sec)
μ	—	Gas viscosity (lb/ft hr)
ϕ	—	Fin efficiency
η	E	Fin effectiveness

9.77 Btu/ft^2 hr °F, fin effectiveness is 82%, fin surface area is 5.23 ft^2/ft, and total surface area is 5.9 ft^2/ft.

Note that the linear velocity V is related to G by the expression

$$V = \frac{G(460 + t)}{144{,}000}$$

```
10 PRINT"HEAT TRANSFER COEFFICIENT IN FINNED TUBES"
15 PRINT"FOR SOLID FINNED TUBES IN STAGGERED ARRANGEMENT"
20 INPUT"GAS FLOW-LB/H,TEMP,TUBE OD-IN,LENGTH-FT,NO OF TUBES WIDE,TR PITCH=";W,T
,D,L,M,X
30 INPUT"FINS/IN,FIB HEIGHT,FIN THK=";N,H,B
40 A=(D/12)+N*B*H/6:G=W/(M*L*((X/12)-A)):V=G*(460+T)/144000!
50 F=3.14*N*(4*D*H+4*H*H+2*B*D+4*B*H)/24
60 S=F+3.14*D*(1-N*B)/12
70 O=.295*G^.681*((1/N)-B)^.313*(.125+.00004*(T-400))/D^.319/H^.2/B^.113
80 E=1/(1+(.33*(O/B)*H*H*((D+2*H)/D)^.5/144))
90 E=1-(1-E)*F/S
100 PRINT" "
110 PRINT"GAS FLOW-LB/H=";W;"GAS TEMP-F=";T
120 PRINT" "
130 PRINT"TUBE OD-IN=";D;"TR PITCH-IN=";X;"TUBE LENGTH-FT=";L
140 PRINT" "
150 PRINT"NO OF TUBES WIDE=";M;"FINS/IN=";N;"FIN HEIGHT=";H
160 PRINT" "
170 PRINT"FIN THK-IN=";B;"GAS MASS VEL-LB/FT2H=";G;"LIN VEL-FT/S=";V
180 PRINT" "
190 PRINT"GAS COEFF-BTU/FT2HF=";O;"FIN EFFECTIVENESS=";E
200 PRINT" "
210 PRINT"FIN AREA-FT2/FT=";F;"TOT SURF AREA-FT2/FT=";S
215 PRINT" "
220 END

HEAT TRANSFER COEFFICIENT IN FINNED TUBES
FOR SOLID FINNED TUBES IN STAGGERED ARRANGEMENT
GAS FLOW-LB/H,TEMP,TUBE OD-IN,LENGTH-FT,NO OF TUBES WIDE,TR PITCH=? 150000,525,3
.5,10.5,12,6
FINS/IN,FIB HEIGHT,FIN THK=? 3,.75,.059                                  " "

GAS FLOW-LB/H= 150000 GAS TEMP-F= 525

TUBE OD-IN= 3.5 TR PITCH-IN= 6 TUBE LENGTH-FT= 10.5

NO OF TUBES WIDE= 12 FINS/IN= 3 FIN HEIGHT= .75

FIN THK-IN= 5.900001E-02 GAS MASS VEL-LB/FT2H= 6393.249 LIN VEL-FT/S= 43.7316

GAS COEFF-BTU/FT2HF= 9.773624 FIN EFFECTIVENESS= .822284

FIN AREA-FT2/FT= 5.23595 TOT SURF AREA-FT2/FT= 5.989681
```

FIG. 3.4 Listing of program 3.4, with results.

Reference

V. Ganapathy, *Applied Heat Transfer*, Pennwell Books, Tulsa, Oklahoma, 1982, p. 500.

PROGRAM 3.5 HEAT TRANSFER COEFFICIENT FOR AIR AND GAS OVER PLAIN TUBES

Input

Gas or air flow
Temperature
Fluid temperature inside tubes
Tube OD
Length of tubes
Number of tubes wide
Transverse pitch
Nature of fluid inside tubes

Output

Air volume
Mass velocity
Density
Heat transfer coefficient

Remarks

The effect of the arrangement, whether it is staggered or inline, is not significant for bare tubes. The nature of the fluid inside the tubes helps estimate the gas film temperature at which gas properties are to be calculated.

The program calculates the heat transfer coefficient for air or flue gases flowing over plain tube bundles arranged in either inline or in staggered fashion, given the gas flow, gas temperature, tube configuration, and whether water and steam or air and gas flows inside the tubes. The results include the gas heat transfer coefficient, gas flow (lb/hr and acfm), linear and mass velocities, and density. There is likely to be a small variation in the heat transfer coefficient due to difference in gas properties between actual and assumed values, but the results are good for engineering purposes.

Theory

The convective heat transfer coefficient when air or gas flows over plain tube bundles may be estimated from the following equation for staggered and inline arrangements.* The difference in heat transfer coefficients between inline and staggered arrangements is not significant.

$$h_c = \frac{0.9G^{0.6}F}{d^{0.4}} \tag{1}$$

where

$$F = \frac{k^{0.67}C_p{}^{0.33}}{\mu^{0.27}} \tag{2}$$

The air and gas properties are estimated at the gas film temperature t_f, as follows. When steam or water flows inside tubes, as in boilers,

$$t_f = 0.5(t_g + t_i)$$

When air or gas flows inside tubes, as in air heaters,

$$t_f = 0.75t_g + 0.25t_i$$

where t_g and t_i are the gas temperature and temperature of fluid inside tubes, respectively.

The factor F may be obtained from air and flue gas from

$$F = 0.094 + 0.00004(t_f - 200) \tag{3}$$

Also

$$G = \frac{12W_g}{N_w L(s_t - d)} \tag{4}$$

$$V = \frac{G}{3600\rho_g} \tag{5}$$

$$\rho_g = \frac{40}{460 + t_g} \tag{6}$$

$$q = \frac{w_g}{60\rho_g} \tag{7}$$

Table 3.5 gives the nomenclature. Figure 3.5 gives the listing as well as the printout.

---

*V. Ganapathy, *Applied Heat Transfer*, Pennwell Books, Tulsa, Oklahoma, 1982.

TABLE 3.5 Nomenclature for Program 3.5

Nomenclature	Program symbol	Description and units
C_p	—	Gas specific heat (Btu/lb °F)
d	D	Tube outer diameter (in.)
F	—	Factor defined in equation (2)
G	G	Gas mass velocity (lb/ft^2 hr)
h_c	H	Convective heat transfer coefficient outside tubes (Btu/ft^2 hr °F)
k	—	Gas thermal conductivity (Btu/ft hr °F)
L	L	Tube length (ft)
N_w	N	Number of tubes wide
q	q	Gas flow (acfm)
s_t	S	Transverse pitch (in.)
t_i	I	Temperature of fluid inside (°F)
t_g	T	Gas temperature outside (°F)
t_f	F	Gas film temperature (°F)
V	V	Gas velocity (ft/sec)
W_g	W	Gas flow (lb/hr)
μ	—	Gas viscosity (lb/ft hr)
ρ_g	R	Gas density (lb/ft^3)

Example

Determine the gas-side convective heat transfer when flue gases flow over air heater tubes with the following parameters:

1. Gas flow = 780,000 lb/hr
2. Average gas temperature = 520°F
3. Average temperature of air inside tubes = 325 °F
4. Tube outer diameter = 2.0 in.
5. Tube length = 23 ft
6. Number of tubes wide = 28
7. Transverse pitch = 4.0 in.
8. Fluid inside tube (air/gas = 1; water/steam = 0) = 1

```
10 PRINT"HEAT TRANSFER -AIR/GAS OVER BARE TUBES"
20 INPUT"GAS FLOW-LB/H,GAS TEMP-F,FLUID TEMP INSIDE TUBES,TUBE OD-IN,TUBE LENGTH
-FT,TUBES WIDE,TR PITCH-IN=";W,T,I,D,L,N,S
30 INPUT"FLUID INSIDE TUBES-INPUT 1 FOR AIR/GAS,0 FOR WATER/STEAM=";J
40 IF J=0 THEN F=.5*(T+I):GOTO 60
50 F=.75*T+.25*I
60 G=12*W/(N*L*(S-D)):H=.9*G^.6*(9.399999E-02+.00004*(F-200))/D^.4
70 R=40/(460+T):Q=W/60/R:V=G/3600/R
80 PRINT" "
90 PRINT"GAS HEAT TRANSFER COEFFICIENT"
100 PRINT" "
110 PRINT"GAS FLOW-LB/H=";W;"FLOW-CFM=";Q;"MASS VELOCITY-LB/FT2H=";G
120 PRINT" "
130 PRINT"GAS LINEAR VELOCITY-FT/S=";V;"GAS TEMP-F=";T;"FLUID TEMP-F=";I
140 PRINT" "
150 PRINT"NO OF TUBES WIDE=";N;"TR PITCH-IN=";S
160 PRINT" "
170 PRINT"GAS DENSITY-LB/CU FT=";R;"GAS HT TR COEFF-BTU/FT2HF=";H
175 PRINT" "
180 END

HEAT TRANSFER -AIR/GAS OVER BARE TUBES
GAS FLOW-LB/H,GAS TEMP-F,FLUID TEMP INSIDE TUBES,TUBE OD-IN,TUBE LENGTH-FT,TUBES
 WIDE,TR PITCH-IN=? 780000,520,325,2,23,28,4
FLUID INSIDE TUBES-INPUT 1 FOR AIR/GAS,0 FOR WATER/STEAM=? 1

GAS HEAT TRANSFER COEFFICIENT

GAS FLOW-LB/H= 780000 FLOW-CFM= 318500 MASS VELOCITY-LB/FT2H= 7267.081

GAS LINEAR VELOCITY-FT/S= 49.45652 GAS TEMP-F= 520 FLUID TEMP-F= 325

NO OF TUBES WIDE= 28 TR PITCH-IN= 4

GAS DENSITY-LB/CU FT= 4.081633E-02 GAS HT TR COEFF-BTU/FT2HF= 14.83252
```

FIG. 3.5 Listing of program 3.5, with results.

Solution

The program is keyed in. In the RUN mode the screen asks for the data in the same order given here and they are fed in. Then the computer solves the various equations and prints out the results.

It is seen that the gas heat transfer coefficient is 14.83 Btu/ft^2 hr °F, gas mass velocity is 7267 lb/ft^2/hr, linear velocity is 49 ft/sec, flow (acfm) is 318,500, and density is 0.04 lb/ft^3.

PROGRAM 3.6 NONLUMINOUS HEAT TRANSFER COEFFICIENT IN TUBE BANKS

Input

Tube OD
Transverse and longitudinal pitch
Gas temperature
Tube surface temperature
Partial pressures of carbon dioxide and water vapor in flue gas

Output

Gas emissivity
Heat flux
Nonluminous heat transfer coefficient

Remarks

If partial pressures of CO_2 and H_2O are not known, use Table 3.6 for good estimates. If a triatomic gas is absent, input 0.001 to execute the program.

At temperatures above 800°F, the nonluminous heat transfer coefficient between flue gas and boiler tubes becomes important, especially if the flue gas contains significant moisture and carbon dioxide. This program predicts the coefficient if the flue gas and tube surface temperatures, tube configuration, and partial pressures of water vapor and carbon dioxide plus sulfur dioxide are known. In case information on the partial pressures of triatomic gases is not available, Table 3.6A may be used for estimates.

Theory

The net interchange of radiation between flue gases and tube bundles is given by the basic equation:*

$$\frac{Q}{A} = \sigma \epsilon_g (T_g{}^4 - T_o{}^4) = h_N (T_g - T_o) \tag{1}$$

The nonluminous heat transfer coefficient h_N may be obtained from the above. The emissivity ϵ_g of the gases is obtained from

*V. Ganapathy, *Applied Heat Transfer*, Pennwell Books, Tulsa, Oklahoma, 1982, p. 475.

TABLE 3.6A Partial Pressures of CO_2 and H_2O[a]

Fuel	p_c	p_w
Natural gas	0.08	0.175
Bituminous coals	0.15	0.075
No. 6 oil	0.12	0.10
No. 2 oil	0.11	0.11
Wood, 30% moisture	0.135	0.195

[a]Combustion calculations based on 20% excess air.

TABLE 3.6B Nomenclature for Program 3.6

Nomenclature	Program symbol	Description and units
A	—	Surface area (ft^2)
d	D	Tube outer diameter (in.)
g	E	Gas emissivity
$c'w$	—	Emissivity due to (CO_2 + SO_2) and H_2O
$'$	—	Correction terms in equation (3)
h_N	H	Nonluminous heat transfer coefficient (Btu/ft_2 hr °F)
K	K	Factor in equation (5)
L	Z	Beam length
q	Q	Heat flux (Btu/ft^2/hr)
Q	—	Duty (Btu/hr)
s_t	S	Transverse pitch (in.)
s_1	L	Longitudinal pitch (in.)
t_g	T	Gas temperature (°F)
t_s	X	Surface temperature (°F)
p_c	C	Partial pressure of (CO_2 + SO_2)
p_w	W	Partial pressure of H_2 (atm)

```
10 PRINT" NON-LUMINOUS HEAT TRANSFER COEFFICIENT IN TUBE BUNDLES"
20 INPUT"TUBE OD,TR PITCH,LONG PITCH(in),GAS TEMP,SURF TEMP OF TUBES(F)";D,S,L,T
,X
30 INPUT"P CO2,PH2O(IF THESE ARE NIL USE 0.00001)=";C,W
40 M=1.08*(S*L-.785*D*D)/D:Z=M/39.36
50 Y=((T-32)/1.8)+273
60 K=(.8+1.6*W)*(1-.00038*Y)*(C+W)/SQR((C+W)*Z)
70 E=.9*(1-EXP(-K*Z))
80 H=.173*E*.9*(((T+460)/100)^4-((X+460)/100)^4)/(T-X)
90 Q=H*(T-X)
95 PRINT" "
100 PRINT"  RESULTS"
105 PRINT" "
110 PRINT"TUBE OD=";D;"TR PITCH=";S;"LONG PITCH=";L
115 PRINT" "
120 PRINT"PARTIAL PRESSURE-H20=";W;"PARTIAL PRESS-CO2=";C
125 PRINT" "
130 PRINT"GAS EMISSIVITY=";E
135 PRINT" "
140 PRINT"GAS TEMP-F=";T;"SURFACE TEMP-F=";X
145 PRINT" "
150 PRINT"NON-LUM HTC-BTU/FT2HF=";H;"HEAT FLUX-BTU/FT2H=";Q
155 PRINT" "
160 END
```

```
 NON-LUMINOUS HEAT TRANSFER COEFFICIENT IN TUBE BUNDLES
TUBE OD,TR PITCH,LONG PITCH(in),GAS TEMP,SURF TEMP OF TUBES(F)? 2,4,3.6,1600,500

P CO2,PH2O(IF THESE ARE NIL USE 0.00001)=? .12,.16
                                "  "
  RESULTS

TUBE OD= 2 TR PITCH= 4 LONG PITCH= 3.6

PARTIAL PRESSURE-H20= .16 PARTIAL PRESS-CO2= .12

GAS EMISSIVITY= .1050696

GAS TEMP-F= 1600 SURFACE TEMP-F= 500

NON-LUM HTC-BTU/FT2HF= 2.551878 HEAT FLUX-BTU/FT2H= 2807.066
```

FIG. 3.6 Listing of program 3.6, with results.

$$\epsilon_g = \epsilon_c + \eta\epsilon_w - \Delta\epsilon \qquad (2)$$

To estimate the emissivity, we need the partial pressures of water vapor, carbon dioxide, and sulfur dioxide and the beam length of the bundle L, which is calculated from

$$L = \frac{1.08(s_t s_l - 0.785d^2)}{d} \qquad (3)$$

Hottel's charts are widely used to determine the emissivity of gases. The following equations, which approximate the charts, give a good estimate of ϵ_g.

$$\epsilon_g = 0.9[1 - \exp(-KL)] \tag{4}$$

where factor K is obtained from

$$K = \frac{(0.8 + 1.6p_w)(1 - 0.38T_g/1000)(p_c + p_w)}{\sqrt{[(p_c + p_w)L]}} \tag{5}$$

where T_g is the gas temperature (Kelvin), p_c and p_w are the partial pressures of carbon dioxide and water vapor (atmospheres), and L is the beam length (meters).

Once h_N is known, the heat flux may be calculated as $q = h_N(t_g - t_s)$. Figure 3.6 gives the listing and printout of results: Table 3.6B gives the nomenclature.

Example

A boiler superheater is made of tube bundle that has 2.0 in. OD tubes at a 4 in. transverse and a 3.6 in. longitudinal pitch. Average gas temperature is 1600°F, and the tube surface temperature is 500°F. (For estimates, fluid temperature plus 50-100°F may be used for surface temperature if a metal temperature determination is not made.) $p_c = 0.12$ and $p_w = 0.16$. Estimate the nonluminous heat transfer coefficient and the heat flux.

Solution

Key in the program. In the RUN mode, the screen asks for the data in the same order given here:

1. Tube OD = 2
2. Transverse pitch = 4.0 in.
3. Longitudinal pitch = 3.6 in.
4. Gas temperature = 1600°F
5. Surface temperature = 500°F
6. Partial pressure of carbon dioxide + sulfur dioxide = 0.12
7. Partial pressure of water vapor = 0.16

Once these data are fed in, the computer goes on to solve all the equations and prints out the results: gas emissivity is 0.105, nonluminous heat transfer coefficient is 2.55 Btu/ft^2 hr °F, and heat flux is 2807 Btu/ft^2 hr. *Caution*: Even if partial pressure of a triatomic gas is zero, input a small number, such as 0.001, to execute the program.

Reference

V. Ganapathy, Estimate nonluminous radiation heat transfer coefficients, *Hydrocarbon Processing*, April 1981, p. 236.

PROGRAM 3.7 ESTIMATING FIN TIP TEMPERATURES*

Input

Tube OD
Tube ID
Fin density
Fin height
Fin thickness
Fin thermal conductivity
Tube length
Number of tubes wide
Transverse pitch
Gas flow
Gas temperature
Tubeside heat transfer coefficient
Tubeside fluid temperature
Fouling factors

Output

Tube wall temperature
Fin tip temperature
Gas heat transfer coefficient
Fin effectiveness
Overall heat transfer coefficient

Remarks

This technique is valid for a staggered tube arrangement and solid fins.

*This program modified from *Oil and Gas Journal*, May 7, 1984, p. 187.

Finned tube bundles are widely used in boilers, heat recovery systems, fired heaters, economizers, and superheaters. Calculation of heat transfer coefficients and fin tip and fin base temperatures involves solution of several equations and is a lengthy procedure.

With the program presented here, one can arrive at the results in a few seconds.

Theory

To arrive at the fin tip or base temperatures, the gas and tubeside heat transfer coefficients must be known.

The gas-side heat transfer coefficient is obtained from the equation of Briggs and Young for solid helical finned tubes in staggered arrangement, which is commonly adopted.*

$$h_g = 0.295G^{0.681}\left(\frac{1}{n} - b\right)^{0.313} \frac{k^{0.67}cp^{0.33}}{d^{0.319}h^{0.2}b^{0.113}\mu^{0.351}} \tag{1}$$

The gas mass velocity G is given by

$$G = \frac{W_g}{N_w L(S_T/12 - A_0)} \tag{2}$$

where the obstruction area A_0 is given

$$A_0 = \frac{d}{12} + \frac{nbh}{6} \tag{3}$$

The outside gas coefficient h_g is to be corrected for fin efficiency η and effectiveness ϕ as

$$\eta = \left[1 - (1 - \phi)\frac{A_f}{A_T}\right] \tag{4}$$

Fin efficiency is obtained from*†

$$\phi = \frac{1}{[1 + 1/3m^2h^2\sqrt{(d + 2h)/d}]} \tag{5}$$

*V. Ganapathy, *Applied Heat Transfer*, Pennwell Books, Tulsa, Oklahoma 1982.

†V. Ganapathy, Evaluating waste heat recovery projects, *Hydrocarbon Processing*, August 1982.

and

$$m = \sqrt{\frac{24h_g}{k_m b}} \tag{6}$$

A_f and A_T, the fin and total surface areas, are obtained from

$$A_f = \frac{\pi n}{24}\,[4dh + rh^2 + 2bd + 4bh] \tag{7}$$

and

$$A_T = A_f + (1 - nb)\frac{\pi d}{12} \tag{8}$$

The overall heat transfer coefficient is then computed from

$$\frac{1}{U} = \left(\frac{1}{h_i} + ff_i\right)\frac{A_T}{A_i} + ff_0 + \frac{1}{\eta h_g} + \frac{A_w}{A_T}\;\frac{d}{24}\,k_m\,\ln\left(\frac{d}{d_i}\right) \tag{9}$$

where ff_i and ff_0 are the fouling factors inside and outside the tubes. The tubeside fouling factor lies in the range of 0.0005 to 0.001 for such fluids as water or steam. The outside fouling factor depends on the type of fuel fired or the type of gas. For the products of combustion from clean fuels, it may be taken as 0.001. For dirty gases, such as the products of combustion of coals or crude oils, it may range from 0.005-0.01.

The tubeside heat transfer coefficient h_i may be estimated from methods described elsewhere. For two-phase flow boiling situations, 1000 Btu/ft^2 hr °F is a conservative estimate.

The thermal conductivity of fins K_m, may be estimated at fluid temperature plus 100°F to start and may be corrected later if necessary. K_m lies in the range 23-28 Btu/ft hr °F for carbon steel fins.

Once U is obtained from heat flux considerations, the fin base temperature t_w and fin tip temperature t_f are found by

$$t_w = t_g - \frac{U(t_g - t_i)}{\eta h_g} \tag{10}$$

$$t_f = t_w + K(t_g - t_w) \tag{11}$$

K is a complicated function of fin efficiency ϕ. However, it may be approximated for engineering purposes by

$$K = (1.42 - 1.4\phi) \tag{12}$$

The program is designed to solve all these equations. The flue gas properties C_p, μ, and k have been considered in the estimation of h_g from equation (1) and it is adequate if gas temperature is fed. Nonluminous radiation coefficients play a small role below 1200°F, and hence their effects were neglected.

An example illustrates the use of the program. The program listing is shown in Fig. 3.7A and the nomenclature in Table 3.7. A typical fin tip configuration is illustrated in Fig. 3.7B.

Example

In a waste heat boiler, 235,000 lb/hr of flue gases at 1000°F transfer energy to steam water mixture at 450°F.

Tube OD: 2.0 in.
Tube ID: 1.78 in.
Fins/in.: 3.0
Fin height: 0.75 in.
Fin thickness: 0.06 in.
Fin thermal conductivity: 24 Btu/ft hr °F
Tube length: 20 ft
Number of tubes wide: 12
Transverse pitch: 4 in.
Gas flow: 235,000 lb/hr
Gas temperature: 1000°F
Inside heat transfer coefficient: 1000 Btu/ft^2 hr °F
Fluid temperature: 450°F
Fouling factor in: 0.001 ft^2 hr °F/Btu
Fouling factor out: 0.002 ft^2 hr °F/Btu

Determine the gas heat transfer coefficients and fin base and tip temperatures.

Solution

Key in the program. The screen asks for all the data items in the same order given here. The computer then solves the equations discussed earlier and prints out the results seen in Fig. 3.7A.

In this example, the fin tip is at 734°F and the base is at 545 °F. Fin effectiveness is 0.74; and the overall heat transfer coefficient is 8.60 Btu/ft^2 hr °F.

TABLE 3.7 Nomenclature for Program 3.7

Nomenclature	Program symbol	Description and units
A_1	P	Fin area (ft^2/ft)
A_0	A	Obstruction area (ft^2/ft)
A_T	P	Total area of finned tube (ft^2/ft)
b	B	Fin thickness (in.)
d	D	Tube OD (in.)
d_1	I	Tube ID (in.)
ff_1	A (30)	Fouling factor inside tubes ($ft^2/hr\ ^\circ F/Btu$)
ff_0	A(40)	Fouling factor outside tubes ($ft^2/hr\ ^\circ F/Btu$)
G	G	Gas mass velocity ($lb/ft^2/hr$)
h	H	Fin height (in.)
h_g	O	Gas-side heat transfer coefficient ($Btu/ft^2\ hr\ ^\circ F$)
h_1	C	Tubeside heat transfer coefficient ($Btu/ft^2/hr\ ^\circ F$)
K	X	Factor used in equation (12)
K_m	K	Fin thermal conduction ($Btu/ft\ hr\ ^\circ F$)
L	L	Length of tube (ft)
m	M	Factor used in equation (6)
n	N	Number of fins per inch
S_T	S	Transverse pitch (in.)
t_1	Q	Tube medium temperature ($^\circ F$)
t_g	T	Gas temperature ($^\circ F$)
t_f	Y	Fin tip temperature ($^\circ F$)
t_w	A(35)	Fin base temperature ($^\circ F$)
U	U	Overall heat transfer coefficient ($Btu/ft^2\ hr\ ^\circ F$)
W_g	W	Gas flow (lb/hr)
η	E	Fin effectiveness, equation (4)
c_p	—	Gas specific heat ($Btu/lb\ ^\circ F$)
μ	—	Viscosity (lb/ft hr)
k	—	Thermal conductivity ($Btu/ft\ hr\ ^\circ F$)

```
1Ø PRINT"FIN TIP TEMPERATURE CALCULATIONS"
2Ø INPUT"TUBE OD,TUBE ID,FINS/IN,FIN HT,FIN THK=";D,I,N,H,B
3Ø INPUT"FIN THERM COND-BTU/FTHF,TUBE LENGTH,NO OF TUBES WIDE,TR PITCH=";K,L,J,S

4Ø INPUT"GAS FLOW,GAS TEMP,TUBE SIDE HTC,TUBE FLUID TEMP=";W,T,C,Q
5Ø INPUT"FOULING FATOR INSIDE,OUTSIDE-FT2HF/BTU=";A3Ø,A4Ø
6Ø A=(D/12)+N*B*H/6:G=W/(J*L*((S/12)-A))
7Ø F=3.14*N*(4*D*H+4*H*H+2*B*D+4*B*H)/24:P=F+3.14*D*(1-N*B)/12
8Ø O=.295*G^.681*((1/N)-B)^.313*(.125+.ØØØØ4*(T-4ØØ))/D^.319/H^.2/B^.113
9Ø M=(24*O/K/B)^.5:E=1/(1+(.33*M*M*H*H*((D+2*H)/D)^.5/144))
1ØØ X=1.42-1.4*E:E=1-(1-E)*F/P:R=.159*LOG(D/I+*P/K/I
11Ø U=(1/(O*E))+(12*P/3.14/I)*((1/C)+A3Ø)+R+A4Ø:U=1/U
12Ø A35=T-U*(T-Q)/E/O:Y=A35+X*(T-A35)
13Ø PRINT" "
14Ø PRINT"FIN TIP TEMPERATURE"
15Ø PRINT" "
16Ø PRINT"GAS FLOW=";W;"GAS TEMP=";T;"TUBE FLUID TEMP=";Q
17Ø PRINT" "
18Ø PRINT"TUBE OD=";D;"TUBE ID=";I;"FINS/IN=";N;"FIN HT=";H;"FIN THK=";B
19Ø PRINT" "
2ØØ PRINT"TUBES WIDE=";J;"LENGTH=";L;"TR PITCH=";S;"SURF AREA-FT2/FT=";P
21Ø PRINT" "
22Ø PRINT"FIN THERML COND=";K;"WALL TEMP=";A35;"FIN TIP TEMP=";Y
23Ø PRINT" "
24Ø PRINT"FOULING FTR IN=";A3Ø;"FOULING FTR OUT=";A4Ø
25Ø PRINT" "
26Ø PRINT"GAS HT TR COEF=";O;"OVERALL HTC=";U;"TUBE SIDE HTC=";C;"FIN EFF=";E
27Ø END

FIN TIP TEMPERATURE CALCULATIONS
TUBE OD,TUBE ID,FINS/IN,FIN HT,FIN THK=? 2,1.78,3,.75,.Ø6
FIN THERM COND-BTU/FTHF,TUBE LENGTH,NO OF TUBES WIDE,TR PITCH=? 24,2Ø,12,4
GAS FLOW,GAS TEMP,TUBE SIDE HTC,TUBE FLUID TEMP=? 235ØØØ,1ØØØ,1ØØØ,45Ø
FOULING FATOR INSIDE,OUTSIDE-FT2HF/BTU=? .Ø21..ØØ2

FIN TIP TEMPERATURE

GAS FLOW= 235ØØØ GAS TEMP= 1ØØØ TUBE FLUID TEMP= 45Ø

TUBE OD= 2 TUBE ID= 1.78 FINS/IN= 3 FIN HT= .75 FIN THK= .Ø6

TUBES WIDE= 12 LENGTH= 2Ø TR PITCH= 4 SURF AREA-FT2/FT= 3.8321Ø9

FIN THERML COND= 24 WALL TEMP= 545.2356 FIN TIP TEMP= 734.7115

FOULING FTR IN= .ØØ1 FOULING FTR OUT= .ØØ2

GAS HT TR COEF= 13.9Ø938 OVERALL HTC= 8.6Ø7362 TUBE SIDE HTC= 1ØØØ FIN EFF=
 .7484Ø83
```

FIG. 3.7A Listing of program 3.7, with results.

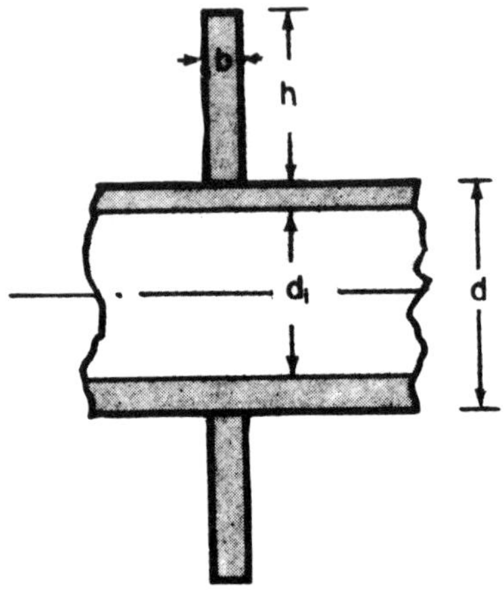

FIG. 3.7B Fin configuration.

PROGRAM 3.8 SUPERHEATER METAL TEMPERATURE CALCULATIONS

Input

Tube OD
Tube ID
Transverse pitch
Longitudinal pitch
Tube length
Number of tubes wide
Row number of tube for which metal temperature is desired
Tube thermal conductivity
Gas flow
Gas temperature
Steam flow in tube
Steam pressure
Steam temperature
Direct radiation from ahead of superheater
Partial pressures of CO_2 and H_2O

Output

Midwall temperature
Surface temperature

Heat flux
Steamside heat transfer coefficient
Nonluminous heat transfer coefficient
Convective heat transfer coefficient
Overall heat transfer coefficient

Remarks

The program considers the effect of direct radiation, nonluminous and convective heat transfer, distribution of external radiation to rows, and nonuniformity around tubes while estimating the wall temperatures.

This program predicts the midwall and outer wall temperature of superheaters and reheaters in boiler plants. Convective as well as nonluminous heat transfer coefficients and direct radiation from cavities are considered which makes the estimate quite accurate. It is assumed that the superheater is of bare tube design, not finned.

Theory

The maximum heat flux at any point on the tube is given by the basic equation*

$$q = \frac{(t_g - t_s)}{(1/(h_c + h_n) + ff_0 + d \ln (d/d_i)/24K_m + d/[d_i(h_i + ff_i)] + (12Q_rF/N_w dL)} \tag{1}$$

where h_c, h_i, and h_n are the convective, steam, and nonluminous heat transfer coefficients, which may be determined by the methods described elsewhere. A nonuniformity factor of 1.45-1.6 is used on h_c.

The direct radiation from the cavity ahead of the superheater Q_r is absorbed in the first few rows of the superheater. The fraction F absorbed depends on the d/s_t ratio*:

$$\phi = \frac{d}{2s_t} - \frac{d}{s_t}\left\{\sin^{-1}\left(\frac{d}{s_t}\right) - \left[\left(\frac{s_t}{d-1}\right)^2 - 1\right]\right\} 0.5 - \frac{s_t}{d} \tag{2}$$

where $F = \phi$ for the first row, $F = [1 - (1 - \phi)]$ for the second row, and so on.

---

*V. Ganapathy, *Applied Heat Transfer*, Pennwell Books, Tulsa, Oklahoma, 1982, pp. 126-130.

The metal conductivity K_m is estimated at the metal temperature of $t_s + 80°F$, which is adequate for engineering estimates.

Once the heat flux q is obtained, the midwall temperature t_m and the outerwall temperature t_0 may be found as

$$t_m = t_s + \frac{qd}{12K_m}\left(\frac{12Km}{h_i d_i} + 0.5 \ln \frac{d + d_i}{2d_i}\right) \tag{3}$$

$$t_0 = t_m + \frac{qd}{24K_m} \ln \frac{2d}{d + d_i} \tag{4}$$

Table 3.8 shows the nomenclature, and Fig. 3.8 gives the listing of the program.

Note that the gas and steam temperature that are input are the temperatures at the location in consideration. For purposes of estimates, this may be obtained by suitably proportioning the temperatures between the inlet and exit of the superheater.

Example

A superheater operates under the following conditions:

1. Tube OD = 2.0 in.
2. Tube ID = 1.686 in.
3. Transverse pitch = 4.0 in.
4. Longitudinal pitch = 6.0 in.
5. Tube length = 32.5 ft
6. Number of tubes wide = 48
7. Row number for metal temperature calculations = 3
8. Tube thermal conductivity (T22 material) = 22 Btu/ft hr °F
9. Gas flow = 375000 lb/hr
10. Gas temperature = 1610°F
11. Steam flow per tube = 6125 lb/hr
12. Steam pressure = 1200 psia
13. Steam temperature = 610°F
14. Direct radiation from cavity = 1.2 MM But/hr
15. Partial pressure of CO_2, SO_2 = 0.12
15. Partial pressure of H_2O = 0.14

Determine the heat flux, heat transfer coefficients, and wall temperatures.

TABLE 3.8 Nomenclature for Program 3.8

Nomenclature	Program symbol	Description and units
d	D	Tube OD (in.)
d_i	I	Tube ID (in.)
ff_i, ff_o	—	Fouling factor inside and outside tubes (ft^2 hr Btu/°F)
F	A30	Fraction of external radiation absorbed
h_c	C	Convective heat transfer coefficient (Btu/ft^2 hr °F)
h_i	A	Steam heat transfer coefficient (Btu/ft^2 hr °F)
h_n	B	Nonluminous heat transfer coefficient (Btu/ft^2 hr °F)
K_m	K	Metal thermal conductivity (Btu/ft hr °F)
L	L	Tube length (ft)
N_w	N	Number of tubes wide
N_r	R	Row number
P	P	Steam pressure (psia)
p_c	A27	Partial pressure of CO_2, SO_2
p_w	A28	Partial pressure of H_2O
q	Q	Maximum heat flux (Btu/ft^2 hr)
Q_r	J	Direct radiation to superheater (Btu/hr)
s_t	S	Transverse pitch (in.)
s_1	X	Longitudinal pitch (in.0
t_m	T	Midwall temperature (°F)
t_o	O	Outer surface temperature (°F)
t_g	M	Gas temperature (°F)
t_s	Y	Steam temperature (°F)
U	U	Overall heat transfer coefficient (Btu/ft^2 hr °F)
W	W	Gas flow (lb/hr)

```
10 PRINT"SUPERHEATER METAL TEMPERATURE"
20 INPUT"TUBE OD,TUBE ID,TR PITCH,LONG PITCH,TUBE LENGTH=";D,I,S,X,L
30 INPUT"NO OF TUBES WIDE,ROW NO,TUB TH COND=";N,R,K
40 INPUT"GAS FLOW,GAS TEMP,STM FLOW/TUBE,STM PRESS-PSIA,STM TEMP=";W,M,V,P,Y
50 INPUT"DIR RADIATION ,P CO2,P H2O(IF PCO2,PH2Ø ARE NIL USE Ø.ØØØ1)=";J,A27,A28

60 GOSUB 5ØØ
70 GOSUB 55Ø
80 GOSUB 6ØØ
92 KF R>2 THEN C=1.45*C ELSE C=1.6*C
1ØØ U=(1/(C+B))+.ØØ5+(D/I)*(1/A)+(D*LOG(D/I)/24/K):U=1/U
11Ø Z=D/S:A3Ø=1.57*Z-Z*(ATN(Z/SQR(1-Z*Z))+SQR((1/Z/Z)-1)-1/Z)
12Ø A31=1:IF R=1 GOTO 16Ø
13Ø A3Ø=(1-A3Ø)*A3Ø:A31=A31+1
14Ø IF A31=R THEN GOTO 16Ø
15Ø GOTO 13Ø
16Ø Q=(M-Y)*U+(18*J/N/D/L)*A3Ø
17Ø T=Y+(Q*D/12/K)*((12*K/A/I)+.5*LOG((D+I)/2/I))
18Ø O=T+(Q*D/24/K)*LOG(2*D/(D+I))
185 J=J/1Ø^6
19Ø PRINT" "
2ØØ PRINT"METAL TEMPERATURE CALCULATIONS"
21Ø PRINT" "
22Ø PRINT"STM PRESS-PSIA=";P,"STM TEMP=";Y;"FLOW/TUBE-LB/H=";V
23Ø PRINT" "
24Ø PRINT"GAS FLOW=";W;"GAS TEMP=";M;"P CO2=";A27;"P H2O=";A28
25Ø PRINT" "
26Ø PRINT"TUBES WIDE=";N;"LENGTH=";L;"TR PITCH=";S;"LONG PITCH=";X
27Ø PRINT" "
28Ø PRINT"TUBE OD=";D;"TUBE ID=";I;"ROW NO=";R;"DIR RAD-MMBTU/H=";J
292 PRINT" "
3ØØ PRINT"CONV COEFF-BTU/FT2HF=";C;"NON-LUM=";B;"STM COEFF=";A
31Ø PRINT" "
32Ø PRINT"OVERALL HTC-BTU/FT2HF=";U;"ROW NO=";R
33Ø PRINT" "
34Ø PRINT"MID WALL TEMP=";T;"OUTER WALL TEMP=";O;"HEAT FLUX-BTU/FT2H=";Q
35Ø PRINT" "
36Ø END
5ØØ G=32*W/(N*L*(S-D)):C=.9*G^(.6)*(9.399999E-Ø2+.ØØØØ4*(.5*(M+Y)-2ØØ))/D^.4
51Ø RETURN
55Ø E=1.Ø8*(S*X-.785*D*D)/D/39.36:A29=((M-32)/1.8)+273
56Ø A3Ø=(.8+1.6*A28)*(1-.ØØØ38*A29)*(A27+A28)/SQR((A27+A28)*E)
57Ø B=.14Ø1*(1-EXP(-A3Ø*E))*1Ø^(-8)*((M+46Ø)^4-(Y+5ØØ)^4)/(M-Y-4Ø)
58Ø RETURN
6ØØ IF Y=Ø THEN GOTO 65Ø
61Ø IF P<2ØØ THEN GOTO 67Ø
62Ø G=1.446*LOG(P)-1Ø:H=.ØØ5*Y-4
63Ø F=.343+.Ø71*G-.Ø43*H-.Ø58*G*H+.Ø28*G*G+.Ø41*H*H
64Ø GOTO 68Ø
65Ø G=.Ø1*P:F=.172+.Ø79*G-.ØØ7297*G*G+.ØØØ257*G^3:Y=115*P^.225
66Ø GOTO 68Ø
67Ø F=.244+.ØØØ132*(Y-327)
68Ø A=2.44*V^.8*F/I^1.8
69Ø RETURN
```

FIG. 3.8 Listing of program 3.8, with results.

```
TUBE OD,TUBE ID,TR PITCH,LONG PITCH,TUBE LENGTH=? 2,1.686,4,6,12.5
NO OF TUBES WIDE,ROW NO,TUB TH COND=? 48,3,22
GAS FLOW,GAS TEMP,STO FLOW/TUBE,STM PRESS-PSIA,STM TEMP=? 375ØØØ,161Ø,6125,12ØØ,
61Ø
DIR RADIATION ,P CO2,P H2O(IF PCO2,PH2Ø ARE NIL USE Ø.ØØØ1)=? 12ØØØØØ,.12,.14

METAL TEMPERATURE CALCULATIONS

STM PRESS-PSIA= 12ØØ          STM TEMP= 61Ø FLOW/TUBE-LB/H= 6125

GAS FLOW= 375ØØØ GAS TEMP= 161Ø P CO2= .12 P H2O= .14

TUBES WIDE= 48 LENGTH= 12.5 TR PITCH= 4 LONG PITCH= 6

TUBE OD= 2 TUBE ID= 1.686 ROW NO= 3 DIR RAD-MMBTU/H= 1.2

CONV COEFF-BTU/FT2HF= 17.98436 NON-LUM= 3.577492 STM COEFF= 463.6714

OVERALL HTC-BTU/FT2HF= 18.32Ø56 ROW NO= 3

MID WALL TEMP= 672.1465 OUTER WALL TEMP= 678.7926 HEAT FLUX-BTU/FT2H= 21462.23
```

FIG. 3.8 (Continued)

Solution

Key in the program. In the RUN mode, the screen asks for the data in the same order given here, and they are fed in. The computer then goes on to calculate the various heat transfer coefficients, heat flux, and wall temperatures. It is seen from the figure that the heat flux is 21,460 Btu/ft^2 hr, steam heat transfer coefficient is 463 Btu/ft^2 hr °F, midwall temperature is 672°F, and outer wall is 679°F. The overall heat transfer coefficient is 18.32 Btu/ft^2 hr °F. Note that the heat transfer coefficient includes a nonuniformity factor to arrive at the maximum heat flux and that the average heat transfer coefficient would be lower than this number.

PROGRAM 3.9 HEAT LOSS FROM MULTILAYERED INSULATED SURFACES

Input

Number of insulation layers
Hot face temperature
Ambient temperature

Wind velocity
Surface emissivity
Thickness and thermal conductivity at any two temperatures for each layer of insulation

Output

Temperature distribution
Heat loss

Remarks

Rapidly converging techniques are used to arrive at the solution. Any number of layers can be handled.

Given data on the insulation, such as the number of layers, thickness, and thermal conductivity, and ambient temperature, hot face temperature, casing emissivity, and wind velocity, this program predicts the temperature profile and the heat loss. Any number of layers can be handled. With thermal conductivity data for each layer, if input at any two known temperatures, the program evaluates the conductivity at the operating temperature. A quick convergence logic is used to speed the iterative procedure.

Theory

Heat loss Q from the surface is given by the basic equation*

$$Q = 0.173\epsilon \times 10^{-8}[(t_s + 460)^4 - (t_a + 460)^4] + 0.296(t_s - t_a)^{1.25}\left(\frac{V + 69}{69}\right)^{0.5} \quad (1)$$

Also

$$Q = \frac{K_m(t_{p+1-i} - t_{p+2-i})}{L_i} \quad (2)$$

Solving for the temperature, we may write from the above

$$t_{p+1-i} = t_{p+2-i} + \frac{QL_i}{K_m} \quad (3)$$

---

*W. C. Turner and J. F. Malloy, *Thermal Insulation Handbook*, McGraw-Hill, New York, 1981, pp. 40-45.

K_m is calculated at the mean layer temperature t_m using a linear relationship

$$K_m = a + bt_m \tag{4}$$

In these equations, i is the layer number. Table 3.9 gives the nomenclature and Fig. 3.9A a typical arrangement; Fig. 3.9B gives the listing of the program with results.

A trial-and-error logic is used to solve for the heat loss and temperature distribution. Briefly, the procedure is as follows.
(1) Obtain the input data, which are

Number of layers
Hot face temperature t_h (°F)
Ambient temperature t_a (°F)
Wind velocity V (ft/min)
Surface emissivity
Thickness of each layer L_i
Thermal conductivity at any two temperatures

(2) Assume the surface temperature $t_s = t_a + 200$. Calculate Q from equation (2).
(3) Assume a reasonable drop of temperature in each layer. To start, it is $X = (t_h - t_s)/p$. Then,

$$t_{p+1-i} = t_{p+2-i} + X$$

(4) Calculate K_m at the average layer temperature of t_m, which is

$$t_m = 0.5(t_{p+1-i} + t_{p+2-i})$$

using equation (4). Coefficients a and b are estimated from the input data on each insulation layer.
(5) Calculate t_{p+1-i} from equation (3) and Q from equation (2).
(6) If t_{p+1-i} from steps 3 and 5 are equal, proceed to step 7. If not correct t_{p+1-i} by averaging the two values and repeat from step 4. (7) Arrive at the hot face temperature in this fashion. Compare this with the hot face temperature given as input. If they are close, within 3 degrees, the results are printed out. The assumed t_s is good. If not, t_s assumed is corrected, and steps 3 to 7 are repeated. To arrive at the correct t_s quickly, the correction factor is halved each time an iteration is made. The first trial value is $t_s \pm 200$. The next trial value is $t_s \pm 100$, and so on, resulting in speedy solution. (8) Results are printed out.

TABLE 3.9 Nomenclature for Program 3.9

Nomenclature	Program symbol	Description and units
ϵ	J	Casing emissivity
K_m	G	Thermal conductivity of layer (Btu in./ft^2 hr °F)
L_i	A(27) to A(26 + p)	Thickness of each layer (in.)
p	P	Total number of layers
Q	Q	Heat loss (Btu/ft^2/hr)
t_a	H	Ambient temperature (°F)
t_h	A	Hot face temperature (°F)
t_s	A (p + 1)	Casing temperature (°F)
t_i	A (2) to A (p)	Intermediate temperatures (°F)
t_m	T	Mean layer temperature (°F)
V	V	Wind velocity (ft/min)

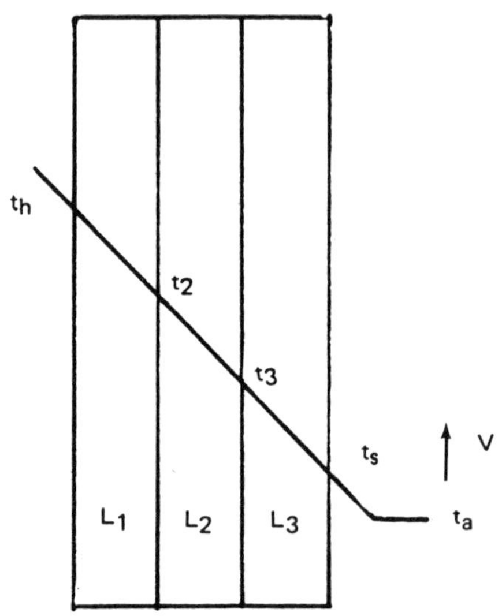

FIG. 3.9A Temperature distribution across insulation layers.

```
10  PRINT"INSULATION PERFORMANCE"
20 DIM A(60)
30 INPUT"number of layers,hot face temp,ambient temp,wind vel in fpm,surf emiss=
";P,A(1),H,V,J
35 PRINT"thermal conductivity data in BTU in/ft2h F"
40 FOR I=1 TO P
50 INPUT "starting from hot face-name,thick,temp 1,cond 1,temp 2,cond 2 =";A$,L,
M,K,N,O
60 A(26+I)=L:A(41+I)=(K-O)/(M-N):A(34+I)=K-A(41+I)*M:A$(I)=A$
70 NEXT I
80 A(P+1)=H+200:U=A(1)
90 Z=A(P+1)-H
100 X=(U-A(P+1))/P
110 W=.01*(A(P+1)+460):Y=.01*(H+460)
120 Q=.173*J*(W^4-Y^4)+93.6*(W-Y)^1.25*SQR((V+69)/69)
130 FOR I=1 TO P
140 A(P+1-I)=A(P+2-I)+X
150 T=.5*(A(P+1-I)+A(P+2-I)):G=A(35+P-I)+A(42+P-I)*T
160 R=A(P+2-I)+Q*A(27+P-I)/G
170 IF ABS(R-A(P+1-I))<3 GOTO 190
180 A(P+1-I)=.5*(R+A(P+1-I)):GOTO 150
190 NEXT I
200 IF ABS(A(1)-U)<3 GOTO 220
210 Z=.5*Z:A(P+1)=A(P+1)+SGN(U-A(1))*Z:GOTO 100
215 PRINT" "
220 PRINT"   RESULTS"
225 PRINT"TEMP -deg F:NAME:THICKNESS,IN"
230 FOR I=1 TO P+1
240 PRINT A(P+2-I),A$(P+2-I),A(28+P-I)
250 NEXT I
255 PRINT" "
260 PRINT"HEAT LOSS -BTU/ft2h=";Q
265 PRINT" "
270 PRINT"AMB TEMP=";H;"WIND VEL-FPM=";V;"EMISS=";J
275 PRINT" "
280 END

INSULATION PERFORMANCE
number of layers,hot face temp,ambient temp,wind vel in fpm,surf emiss=? 2,1800,
80,440,.9
thermal conductivity data in BTU in/ft2h F
starting from hot face-name,thick,temp 1,cond 1,temp 2,cond 2 =? lay 1,4,1600,1.
7,1400,1.4
starting from hot face-name,thick,temp 1,cond 1,temp 2,cond 2 =? lay 2,4,1000,.9
,600,.5
   RESULTS
TEMP -deg F:NAME:THICKNESS,IN
 137.0313                      0
 1338.142     lay 2            4
 1802.481     lay 1            4

HEAT LOSS -BTU/ft2h= 191.4249

AMB TEMP= 80 WIND VEL-FPM= 440 EMISS= .9
```

FIG. 3.9B Listing of program 3.9, with results. (*Source*: *Oil and Gas Journal*, Aug. 19, 1985, p. 125.)

Example

A fired heater has two layers of insulation (ceramic fiber):

Number of layers = 2
Hot face temperature = 1800°F
Ambient temperature = 80°F
Wind velocity = 440 ft/min (5 mph)
Surface emissivity = 0.9
Thickness of layer 1 (starting from hot face) = 4 in.
Thermal conductivity at 1600°F = 1.7 Btu in./ft^2 hr °F
Thermal conductivity at 1400°F = 1.4 Btu in./ft^2 hr °F
Thickness of layer 2 = 4 in.
Thermal conductivity at 1000°F = 0.9 Btu in./ft^2 hr °F
Thermal conductivity at 600°F = 0.5 Btu in./ft^2 hr °F

Determine the temperature distribution and heat loss from the surface.

Solution

Key in the program. In the RUN mode, the screen asks for the data in the same order given here, and they are fed in. The computer goes through the iterative logic and arrives at the final results. From Fig. 3.9B it is seen that the heat loss is 191 Btu/ft^2 hr, the intermediate layer temperature is 1338°F, and the casing temperature is 137°F. Due to the error margin used for iteration, the hot face temperature printed out will be close to the input data but may not exactly match it.

PROGRAM 3.10 SIZING INSULATION TO AVOID SURFACE CONDENSATION*

Input

Fluid temperature inside the pipe
Ambient temperature
Relative humidity
Wind velocity
Pipe diameter

---

*This program modified from *Heating, Piping, and Air Conditioning*, August 1984, p. 102.

Temperature and conductivity of insulation at any two temperatures
Casing emissivity

Output

Surface temperature
Heat loss
Thickness of insulation required to exceed dew point
Dew point of water vapor

Remarks

If fluid temperature is greater than the dew point, then the computer prints out the fact that condensation is not possible. A trial and error procedure that converges quickly is adopted to arrive at the minimum thickness. Fluid and ambient temperatures must be positive.

One of the major problems in process plants handling cold fluids is the sizing of insulation to avoid surface condensation of water vapor. This situation occurs when the casing temperature of insulation drops below the water dew point corresponding to ambient conditions.

With the program shown in Fig. 3.10A the solution can be arrived at in a few seconds.

Theory

Insulation surface temperature calculations involve a trial-and-error procedure.

To solve for T_s, we assume a given value and calculate Q from equations (1) and (2) (Table 3.10A). If the values for Q do not agree, we change our T_s value until they do. To arrive quickly at the correct T_s value, the first trial value is 0.5 $(T_i + T_a)$. The next trial value is this number plus or minus half the value, depending on whether the assumed T_s is assumed to be smaller or larger. Once T_s is calculated for a given thickness of insulation, we have to check if it is greater than the water dew point. Dew point is a function of saturated vapor pressure. The following equation is derived from steam tables:

$$SVP = 0.08 + 281 \times 10^{-9} T_a^{3.25}$$

```
5 CLS:KEY OFF:CLEAR
10 PRINT"INSULATION THICKNESS TO AVOID CONDENSATION":PRINT" "
15 PRINT"PROGRAM NOT APPLICABLE FOR TEMPERATURES BELOW 0 F":PRINT" "
20 INPUT"FLUID TEMP,AMBIENT TEMP,REL HUM-%,WIND VEL-FPM=";A,G,H,V:PRINT" "
25 INPUT"PIPE OD,TEMP 1,COND 1,TEMP 2,COND 2,EMISS=";D,B,M,C,N,E:PRINT" "
30 T=.25:TT=A:GOSUB 1000
35 SVP1=SVP:TT=G:GOSUB 1000
40 SVP2=SVP*.01*H:IF SVP1>=SVP2 THEN PRINT"NO CONDENSATION POSSIBLE AS DEW POINT
 IS BELOW OR CLOSE TO FLUID TEMP":GOTO 240
50 GOSUB 500
55 IF ABS(X-G)<2 THEN PRINT"DEW POINT IS VERY CLOSE TO AMBIENT TEMP-LARGE THICKN
ESS WOULD BE NEEDED":GOTO 240
60 U=.5*(G+A):Z=U-A
70 Q=.173*E*(((G+460)/100)^4-((U+460)/100)^4)+.296*(G-U)^1.25*((V+69)/69)^.5
80 K=M-((N-M)/(C-B))*(B-.5*(U+A)):R=((D+2*T)/(2*K))*LOG((D+2*T)/D):S=(U-A)/R
110 IF ABS((Q-S)/Q)<.02 THEN 130
120 Z=.5*Z:U=U+SGN(Q-S)*Z:GOTO 70
130 IF (U-X)>1 THEN 150
140 T=T+.25:GOTO 60
150 PRINT" "
160 PRINT"INSULATION CALCULATIONS":PRINT" "
170 PRINT"AMB TEM=";G;"FLUID TEMP=";A;"WIND VEL-FPM=";V;"EMISS=";E:PRINT" "
190 PRINT"REL HUM-%=";H;"DEW POINT-WATER=";X;"PIPE OD=";D:PRINT" "
200 PRINT"INSUN THK=";T;"SURFACE TEMP=";U;"HEAT LOSS-BTU/FT2H=";Q:PRINT" "
240 END
500 IF SVP2>=.088 THEN GOTO 550
510 Y=20:RA=19
520 SVP=.01895+6.1759E-04*Y+4.9051E-05*Y*Y
530 IF ABS((SVP-SVP2)/SVP2)<.05 THEN 545
540 RA=.5*RA:Y=Y+SGN(SVP2-SVP)*RA:GOTO 520
545 X=Y:GOTO 560
550 X=((SVP2-.08)/(281*10^(-9)))^(1/3.25)
560 RETURN
1000 IF TT<40 THEN SVP=.01895+6.1759E-04*TT+4.9051E-05*TT*TT:GOTO 1020
1010 SVP=.08+281*10^(-9)*TT^3.25
1020 RETURN

INSULATION THICKNESS TO AVOID CONDENSATION

PROGRAM NOT APPLICABLE FOR TEMPERATURES BELOW 0 F

FLUID TEMP,AMBIENT TEMP,REL HUM-%,WIND VEL-FPM=? 15,100,75,220

PIPE OD,TEMP 1,COND 1,TEMP 2,COND 2,EMISS=? 6.625,100,.26,200,.3,.1

INSULATION CALCULATIONS

AMB TEM= 100 FLUID TEMP= 15 WIND VEL-FPM= 220 EMISS= .1

REL HUM-%= 75 DEW POINT-WATER= 90.67462 PIPE OD= 6.625001

INSUN THK= 1.75 SURFACE TEMP= 92.36328 HEAT LOSS-BTU/FT2H= 8.59966
```

FIG. 3.10A Listing of program 3.10, with results.

Dew point is the temperature corresponding to a saturated vapor pressure of RH × SVP, where RH is the relative humidity expressed as a fraction. If the surface temperature calculated is found to be more than the dew point, the results are printed out; otherwise, the insulation thickness is increased by 0.5 in., and the whole process is repeated until $T_s > T_d$. Figure 3.10B shows the logic used.

TABLE 3.10A Equations Used in Program 3.10

The heat flow from the ambient to the insulation or the pipe is

$$Q_1 = 0.173\epsilon \left[\left(\frac{T_a + 460}{100}\right)^4 - \left(\frac{T_s + 460}{100}\right)^4\right] + 0.296\,(T_a - T_s)^{1.25}\sqrt{\frac{(V + 69)}{69}} \qquad (1)$$

The heat flow from the surface of the insulation to the fluid is

$$Q_2 = (T_s - T_i)/R \qquad (2)$$

R, the thermal resistance of the insulation, is given by:

$$R = \frac{(D + 27)}{2K}\ln[(D + 2T)/D] \qquad (3)$$

Table 3.10B shows the nomenclature and symbols used in the program. Figure 3.10 shows the results for the following example.

Example

A 6 in. schedule 40 pipe handles a fluid at 15°F. The ambient temperature is 100°F, relative humidity is 75%, and wind velocity is 220 fpm.

The insulation used has thermal conductivities of 0.26 and 0.30 Btu hr in./ft^2 °F at 100 and 200°F, respectively. The casing has an emissivity of 0.1. Determine the thickness of insulation to be used, the corresponding surface temperature, and the heat loss.

Solution

The window display asks for the following data in order (the inputs are shown in parentheses following the input titles):

Fluid temperature (15)
Ambient temperature (100)
Relative humidity (75)
Wind velocity (220)
Pipe diameter (6.625)
Temperature 1 (100)

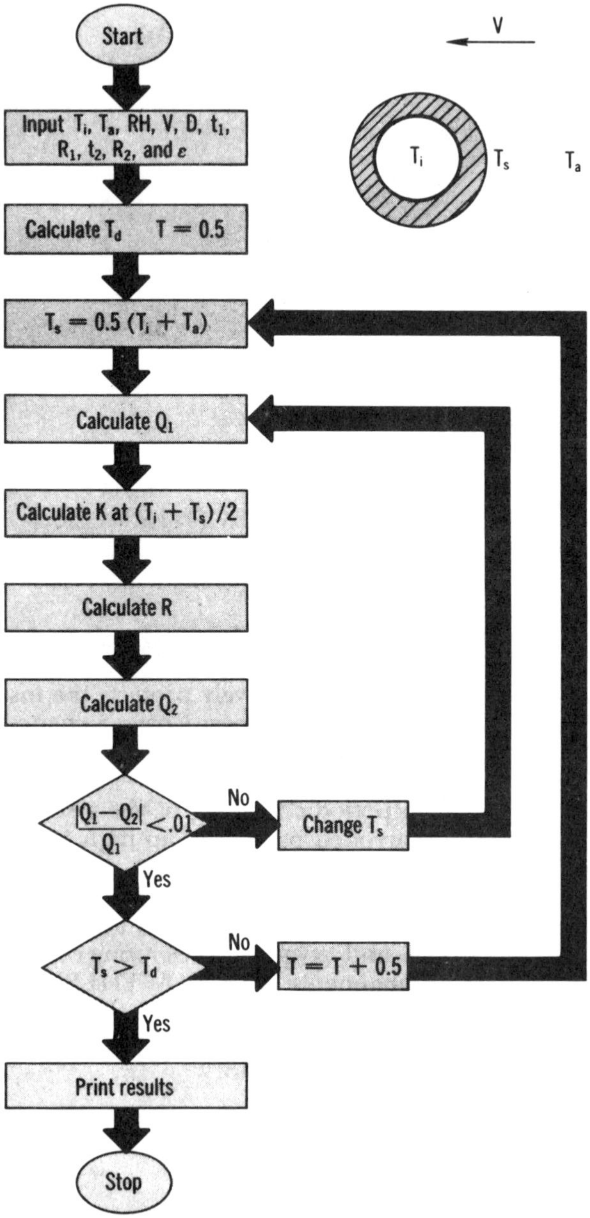

FIG. 3.10B Logic used to solve for T, Q, and T_S.

TABLE 3.10B Nomenclature and Symbols Used in Program 3.10

Nomenclature	Program symbol	Description and units
D	D	Pipe outer diameter (in.)
ϵ	E	Casing emissivity (0.1 for aluminum)
k_1, k_2	M,N	Thermal conductivity of insulation at temperatures t_1 and t_2 (Btu hr in./ft^2/°F)
K	K	Thermal conductivity of insulation at $(T_i + T_s)/2$
Q_1, Q_2	Q,S	Heat loss given by equations (1) and (2)
RH	R	Relative humidity
SVP	—	Saturated vapor pressure (psia)
t_1, t_2	B,C	Temperatures at which insulation conductivity is input (any two values) (°F)
T	T	Insulation thickness (in.)
T_1	A	Fluid temperature (°F)
T_d	X	Water dew point (°F)
T_a	G	Ambient temperature (°F)
T_s	U	Surface temperature of insulation (°F)
V	V	Wind velocity (fpm)
R	R	Thermal resistance of insulation (ft^2 °F/Btu hr)

Conductivity 1 (0.26)
Temperature 2 (200)
Condition 2 (0.30)
Emissivity (0.1)

The computer then solves for T_s for an initial thickness of 0.5 in., compares T_s with T_d, and increases the thickness by another 0.25 in. until $T_s > T_d$. The results are printed out as shown in Fig. 3.10A.

Bibliography

Turner and Malloy, *Handbook of Insulation*, McGraw-Hill, 1982.

4

Steam Utilization

PROGRAM 4.1 STEAM PROPERTIES AFTER EXPANSION

Input

Inlet steam pressure
Temperature
Exit pressure
Efficiency of expansion

Output

Enthalpy of steam at inlet
Enthalpy at exit
Steam temperature at inlet
Steam temperature at exit
Steam quality at exit

Remarks

If steam at inlet is in a saturated condition, input 0 for temperature and the program calculates the value. ASME formulations are used for estimating steam properties. The results are helpful in cogeneration studies.

Engineers involved in cogeneration projects and steam cycle analysis often have to calculate the enthalpy, quality, and temperature of steam after expansion in a steam turbine from an initial pressure P_1 to final pressure P_2. This calculation appears to be simple, but in reality is tedious, as several checks and double interpolations have to be performed if steam tables are used. The initial condition of steam with this program can be either saturated or superheated.

Theory

Steam at pressure P_1 and temperature t_1 is expanded to pressure P_2 at an efficiency E defined as

$$E = \frac{h_1 - h_2}{h_1 - h_{2s}} \qquad (1)$$

where h_{2s} is the enthalpy corresponding to isentropic conditions, that is, when entropy s_1 at P_1 is equal to that at P_2.

If the initial steam is in a saturated condition, input of t_1 = 0 will accomplish estimation of t_1 using the equation

$$t = 115P^{0.225} \tag{2}$$

The isentropic process is then established. At pressure P_2 we have entropy $s_{2s} = s_1$. A check is made to see the steam is wet or superheated. If wet, the quality x is obtained from

$$x = \frac{s_{2s} - s_{2f}}{s_{2g} - s_{2f}} \tag{3}$$

The isentropic enthalpy h_{2s} is computed at P_2.

$$h_{2s} = xh_{2g} + (1 - x)h_{2f} \tag{4}$$

If steam at P_2 is not wet, that is, when $s_{2s} > s_{2g}$, a separate subroutine evaluates t_{2s} at P_2 corresponding to s_{2s}.

Once t_{2s} is known, h_{2s} is obtained. Now, using equation (1), the final enthalpy h_2 is evaluated. Using an iterative loop, t_2 is obtained. If the steam is wet, the quality x is determined from enthalpy calculations:

$$x = \frac{h_2 - h_{2f}}{h_{2g} - h_{2f}} \tag{5}$$

Figure 4.1A shows the flow diagram and Table 4.1 the nomenclature. Quick convergence techniques were used to arrive at the correct solution. ASME 1967 equations, described in the Appendix, were used in steam property evaluations. Two examples illustrate the use of the program, a listing of which is given in Fig. 4.1B. One is for saturated steam and the other for superheated steam.

Example 1

In a cogeneration plant, steam at 800 psia and at 900° F is expanded to 250 psia with an efficiency of 70%. Calculate

1. Initial enthalpy
2. Final enthalpy
3. Final quality
4. Final temperature

Solution

Key in the program. In the RUN mode, the screen asks for the data in the same order given here. Once the data are fed in, the computer goes on to perform the various checks and iterations and

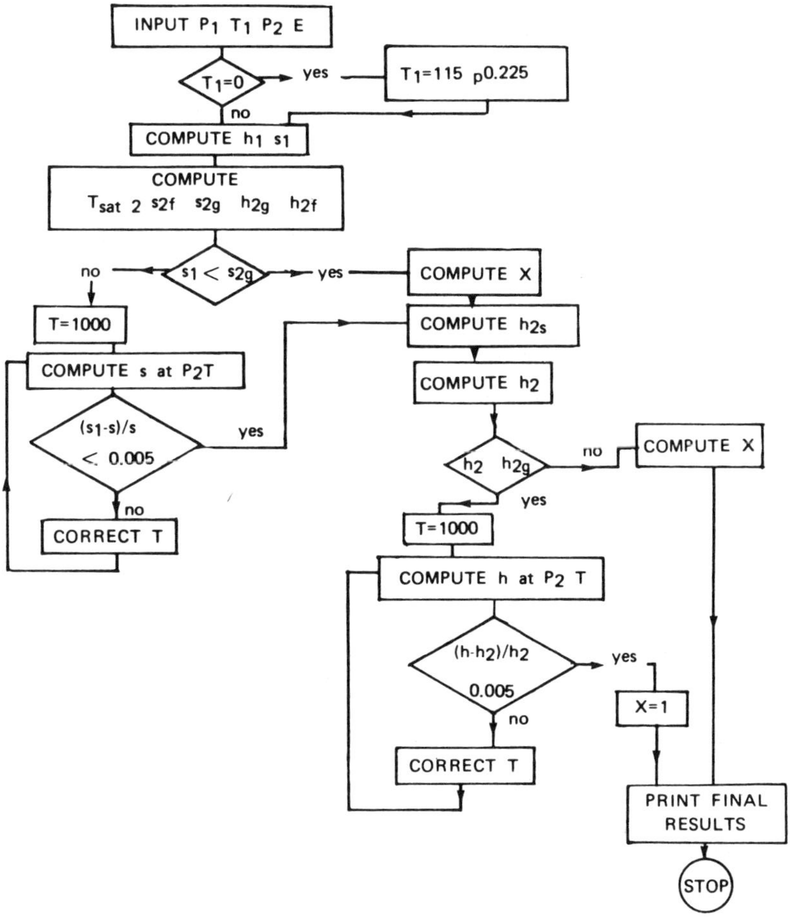

FIG. 4.1A Flow diagram for calculation of steam properties after expansion.

finally prints out the results. It is seen from Fig. 4.1B that exit steam is in a superheated condition (quality is 1), inlet enthalpy is 1457, and final enthalpy is 1356 Btu/lb. Final temperature is 661 °F.

Example 2

This is the same case as example 1 except that the steam is in a saturated condition.

```
10 PRINT"steam properties after expansion"
15 PRINT"IF INITIAL STEAM IS SATURATED INPUT 0 FOR TEMPERATURE"
20 INPUT"inlet press-psia,temperature-f,exit press-psia,efficiency of expn %=";A
,B,C,A32
30 IF B>0 THEN 50
40 B=115*A^.225
50 P=A:T=B:GOSUB 500:E=Z:A27=S
60 I=115*C^.225:T=I:GOSUB 400
70 IF A27>G THEN 100
80 X=(A27-H)/(G-H):F=X*A28+(1-X)*A29:T=I
90 F=E-.01*A32*(E-F):GOTO 140
100 X=1:A31=1000-I:T=1000:P=C
105 GOSUB 500:A30=S
110 IF ABS ((A27-A30)/A30)<.003 THEN 130
120 A31=.5*A31:T=T+SGN(A27-A30)*A31:GOTO 105
130 GOSUB 500:F=Z:F=E-.01*A32*(E-F)
140 IF F>A28 THEN 160
150 X=(F-A29)/(A28-A29):T=I:GOTO 200
160 X=1:P=C:A31=1000-I:T=1000
170 GOSUB 500
180 IF ABS((F-Z)/Z)<.003 THEN 200
190 A31=.5*A31:T=T+SGN(F-Z)*A31:GOTO 170
200 PRINT"RESULTS-EXPANSION OF STEAM"
210 PRINT" "
220 PRINT"STM PRESS IN-PSIA=";A;"STM TEMP IN-F=";B;"ENTH IN-BTU/LB=";E
225 PRINT" "
230 PRINT"STM PRES-OUT-PSIA=";C;"STM TEMP OUT-F=";T,"QUALITY=";X
235 PRINT" "
240 PRINT"EFFICIENCY OF EXPN %=";A32;"ENTH OUT-BTU/LB=";F
245 PRINT" "
250 END
400 T=.01*T:H=(.000358*T^4-.0404*T^3+.0809*T*T+1.595*T-.325)*.1
410 G=1+((-.0337*T^3+.4778*T*T-3.192*T+12.571)/10)
420 A28=1031+113.3*T-45.4*T*T+10.8*T^3-.959*T^4
430 IF T<2.4 THEN 450
440 A29=427.5-389.7*T+187.8*T*T-30.9*T^3+1.88*T^4:GOTO 460
450 A29=100*T-32
460 T=100*T:RETURN
500 T=273.1+(T-32)/1.8:P=P/14.696
510 K=80870!/T/T:L=10^K*(-2641.62/T):M=1.89+L:N=M*P*P/T/T
520 O=2+(372420!/T/T):Q=O*L:R=1.89+Q:U=(.21828*T-126970!/T):V=2*U*R-(M/T)*126460
!
530 W=82.54-162460!/T:Y=2*W*R-(M/T)*162460!
550 Z=775.6+.63596*T+1.62467E-04*T*T+20.5697*LOG(T)
560 Z=Z+.043557*(R*P+.5*N*(Y+M*(W+V*N)))
570 S=(((U*M-2*V)*.5*M*N-Y)*.5*N+(M-R)*P)/T*(-.0241983)-.355579-11.4276/T
580 S=S+1.8052E-04*T-.11022*LOG(P)+.35164*LOG(T)
590 P=14.696*P:T=(T-273.15)*1.8+32
600 RETURN
```

FIG. 4.1B Listing of program 4.1, with results.

```
RESULTS-EXPANSION OF STEAM

STM PRESS IN-PSIA= 800 STM TEMP IN-F= 900 ENTH IN-BTU/LB= 1457.63

STM PRES-OUT-PSIA= 250 STM TEMP OUT-F= 661.1062          QUALITY= 1

EFFICIENCY OF EXPN %= 70 ENTH OUT-BTU/LB= 1356.105


steam properties after expansion
IF INITIAL STEAM IS SATURATED INPUT 0 FOR TEMPERATURE
inlet press-psia,temperature-f,exit press-psia,efficiency of expn %=? 800,0,250,
70
RESULTS-EXPANSION OF STEAM

STM PRESS IN-PSIA= 800 STM TEMP IN-F= 517.4794 ENTH IN-BTU/LB= 1199.652

STM PRES-OUT-PSIA= 250 STM TEMP OUT-F= 398.322          QUALITY= .9210682

EFFICIENCY OF EXPN %= 70 ENTH OUT-BTU/LB= 1137.769
```

FIG. 4.1B (Continued)

TABLE 4.1 Nomenclature for Program 4.1

Nomenclature	Program symbol	Description and units
E	A32	Efficiency of expansion (%)
h_1	E	Initial enthalpy (Btu/lb)
h_2	F	Final enthalpy (Btu/lb)
h_{2f}	A29	Saturated liquid enthalpy at P_2 (Btu/lb)
h_{2g}	A28	Saturated steam enthalpy at P_2 (Btu/lb)
P_1	A	Initial pressure (psia)
p_2	C	Final pressure (psia)
s_1	A37	Initial entropy (Btu/lb) R
s_{2f}	H	Saturated liquid entropy at P_2
s_{2g}	G	Saturated vapor enthalpy at P_2
t_1	B	Initial steam temperature (°F)
t_2	T	Final steam temperature (°F)
x	X	Steam quality (fraction)

Solution

The same data are keyed in, except that initial steam temperature is fed in as 0. The program calculates the saturation temperature and does the rest of the calculations as before. It is seen that the initial temperature and enthalpy are 517°F and 1199 Btut/lb; final temperature and enthalpy are 398°F and 1137 Btu/lb. Final steam is wet, with quality of 0.92.

Note: If isentropic conditions are to be evaluated, use E = 100%.

PROGRAM 4.2 STEAM PROPERTIES RELATED TO THROTTLING OF STEAM

Input

Steam inlet pressure
Temperature
Quality
Exit pressure

Output

Steam enthalpy
Exit quality
Exit temperature
Steam volume at inlet
Steam volume at exit

Remarks

Saturated steam temperature at the inlet is calculated by the program if 0 is input for temperature. The results are helpful in predicting downstream conditions in control or safety valves.

Steam is throttled in several applications in process and power plants when it is pressure reduced, for instance in control or safety valves. Often, it is required to predict the state of steam after this process, which is isenthalpic, that is, enthalpy remains constant. The final steam could be wet, dry, or superheated.

The estimation of specific volume v_2, temperature t_2, and quality x_2 after throttling from initial conditions of pressure P_1, temperature t_1, and quality is tedious, involving several double

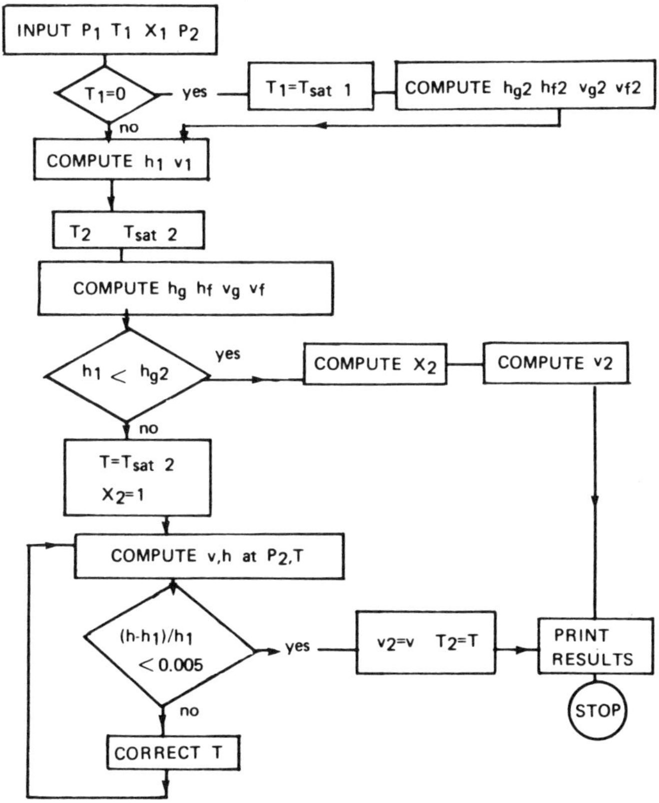

FIG. 4.2A Logic diagram for isenthalpic process calculations.

interpolations if a steam table is used. Figure 4.2A shows the logic used.

With this program, one can handle initially wet, dry, or superheated steam. If the steam is wet or saturated, steam temperature is estimated using the equation

$$t = 115P^{0.225}$$

For two examples, ASME 1967 equations were used for estimating the steam properties. Table 4.2 shows the nomenclature; Fig. 4.2B gives the listing along with the printout of results.

TABLE 4.2 Nomenclature for Program 4.2

Nomenclature	Program symbol	Description and units
h	E	Enthalpy (Btu/lb)
h_f	A29	Saturated water enthalpy (Btu/lb)
h_g	A28	Saturated steam enthalpy (Btu/lb)
P_1	A	Initial pressure (psia)
P_2	C	Final pressure (psia)
t_1	B	Initial temperature (°F)
t_2	T	Final temperature (°F)
v_1	A30	Initial specific volume (ft^3/lb)
v_2	F	Final specific volume (ft^3/lb)
v_f	—	Specific volume of saturated liquid (ft^3/lb)
v_g	—	Specific volume of saturated steam (ft^3/lb)
x_1	X	Initial quality (fraction)
x_2	H	Final quality

```
ISENTHALPIC PROCESS-RESULTS

INLET PRESS-PSIA= 1000 TEMP-F= 900 QUALITY= 1 SP VOL-CU FT/LB= .7595135

ENTHALPY-BTU/LB= 1450.375

EXIT PRESS-PSIA= 500 TEMP-F= 866.1174 QUALITY= 1 SP VOL-CU FT/LB= 1.525889


STEAM PROPERTIES AFTER THROTTLING
STEAM PRESSIN-PSIA,STEAM TEMP(IF SATURATED INPUT 0),QUALITY-FRACTION,EXIT PRESSU
RE-PSIA=? 500,0,.8,15
ISENTHALPIC PROCESS-RESULTS

INLET PRESS-PSIA= 500 TEMP-F= 465.5497 QUALITY= .8 SP VOL-CU FT/LB= .7482202

ENTHALPY-BTU/LB= 1053.447

EXIT PRESS-PSIA= 15 TEMP-F= 211.4139 QUALITY= .8994999 SP VOL-CU FT/LB=
 25.89042
```

FIG. 4.2B Listing of program 4.2, with results.

```
10 PRINT"STEAM PROPERTIES AFTER THROTTLING"
20 INPUT"STEAM PRESSIN-PSIA,STEAM TEMP(IF SATURATED INPUT 0),QUALITY-FRACTION,EX
IT PRESSURE-PSIA=";A,B,X,C
30 IF B=0 GOTO 50
40 P=A:T=B:GOSUB 500:E=Z:A30=F:GOTO 110
50 T=115*A^.225:B=T:GOSUB 400
55 P=A:GOSUB 500:A28=Z
60 E=X*A28+(1-X)*A29:GOSUB 650
65 A30=X*F+(1-X)*A27
70 T=115*C^.225:GOSUB 400
75 P=C:GOSUB 500:A28=Z
80 IF E>A28 GOTO 110
90 H=(E-A29)/(A28-A29):GOSUB 650
100 F=H*F+(1-H)*A27:GOTO 150
110 I=115*C^.225:T=1000:G=1000-I:P=C:H=1
120 GOSUB 500
130 IF ABS((E-Z)/Z)<.005 GOTO 145
140 G=.5*G:T=T+SGN(E-Z)*G:GOTO 120
145 PRINT" "
150 PRINT"ISENTHALPIC PROCESS-RESULTS"
160 PRINT" "
170 PRINT"INLET PRESS-PSIA=";A;"TEMP-F=";B;"QUALITY=";X;"SP VOL-CU FT/1B=";A30
175 PRINT" "
180 PRINT"ENTHALPY-BTU/LB=";E
185 PRINT" "
190 PRINT"EXIT PRESS-PSIA=";C;"TEMP-F=";T;"QUALITY=";H;"SP VOL-CU FT/1B=";F
200 END
400 T=.01*T:IF T<2.4 THEN 420
410 A29=427.5-389.7*T+187.8*T*T-30.9*T^3+1.88*T^4:GOTO 430
420 A29=100*T-32
430 T=100*T:RETURN
500 T=273.1+(T-32)/1.8:P=P/14.696
510 K=80870!/T/T:L=10^K*(-2641.62/T):M=1.89+L:N=M*P*P/T/T
520 O=2+(372420!/T/T):Q=O*L:R=1.89+Q:U=(.21828*T-126970!/T):V=2*U*R-(M/T)*126460
!
530 W=82.54-162460!/T:Y=2*W*R-(M/T)*162460!
540 F=(((U*M*N+W)*N/P+1)*M+4.5504*T/P)*.0160185
550 Z=775.6+.63596*T+1.62467E-04*T*T+20.5697*LOG(T)
560 Z=Z+.043557*(R*P+.5*N*(Y+M*(W+V*N)))
570 S=(((U*M-2*V)*.5*M*N-Y)*.5*N+(M-R)*P)/T*(-.0241983)-.355579-11.4276/T
580 S=S+1.8052E-04*T-.11022*LOG(P)+.35164*LOG(T)
590 P=14.696*P:T=(T-273.15)*1.8+32
600 RETURN
650 T=.01*T:A27=1/(-.1006*T*T*T+.4743*T*T-3.049*T+64.912)
660 IF T>4.6 THEN 700
670 IF T>3 THEN 690
680 F=80.82*T^4-767.06*T^3+2726.21*T*T-4337.7*T+2647.9:GOTO 710
690 F=.529*T^4-9.208*T^3+60.986*T*T-183.361*T+213.456:GOTO 710
700 F=.02097*T^4-.542*T^3+5.319*T*T-23.676*T+40.709
710 T=100*T:RETURN
```

FIG. 4.2B (Continued)

Example 1

Superheated steam at 1000 psia and 900° F is pressure reduced to 500 psia in a control valve. Estimate

1. Enthalpy of steam
2. Initial and final specific volume
3. Initial and final steam temperature
4. Initial and final steam quality

Solution

Key in the program. In the RUN mode, the screen asks for the inlet pressure (1000), inlet temperature (900), initial quality (1), and final pressure (500), which are fed in. If initial steam is in a saturated condition, input 0 for temperature, and the program will compute saturation temperature.

Once the data are fed in, the computer goes on to perform the various calculations and checks per the flowchart and finally prints out the results. It is seen that the enthalpy is 1450 Btu/lb, final temperature is 866° F, and initial and final specific volumes are 0.7595 and 1.525 ft^3/lb.

Example 2

Wet steam at 80% quality and at 500 psia is discharged to the atmosphere in a safety valve. Determine the various parameters.

Solution

Input the various data. Quality is 0.8, and steam temperature is fed in as 0. Final pressure is 15 psia. The computer goes on to perform the various checks and prints out the results. It is seen that enthalpy is 1053 Btu/lb, initial and final specific volumes are 0.748 and 25.891 ft^3/lb, and initial and final steam qualities are 0.8 and 0.899, respectively.

PROGRAM 4.3 WATER REQUIRED FOR DESUPERHEATING STEAM

Input

Steam pressure
Initial and final steam temperatures
Water temperature

Output

Enthalpy at inlet
Enthalpy at exit
Final steam temperature
Water enthalpy
Water-final steam ratio

Remarks

Final steam can be in saturated condition, when the input of 0 for temperature results in the estimation of saturation temperature. ASME 1967 formulations are used for steam property estimation.

Steam temperature control in process and power plants is often accomplished by injecting a spray of water into steam using a desuperheater, as shown in Fig. 4.3A. Given the initial and final steam conditions, the program computes the enthalpies of steam and the water-steam ratio for achieving the desired final steam temperature. ASME 1967 formulations as given in the Appendix are used for determining the steam properties.

Theory

From an energy balance across the desuperheater, we have

$$W_1 h_1 + W_f h_f = (W_1 + W_f) h_2 = W_s h_2 \tag{1}$$

Simplifying, we have

$$\frac{W_f}{W_s} = \frac{h_1 - h_2}{h_1 - h_f} \tag{2}$$

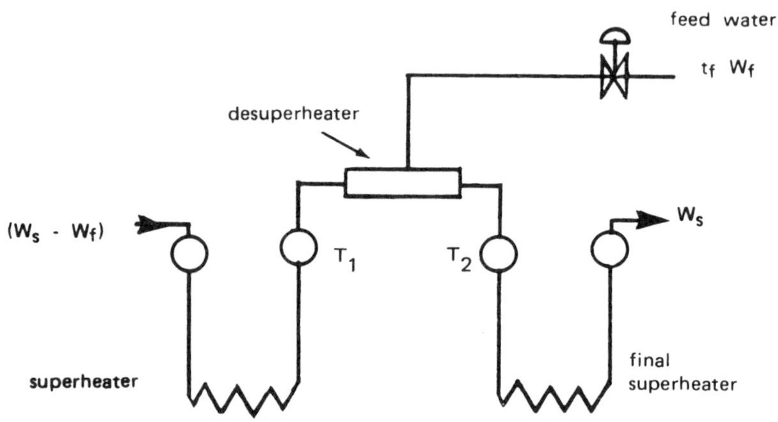

FIG. 4.3A Steam temperature control.

Table 4.3 gives the nomenclature.

The pressure drop across the desuperheater is usually very small and can be neglected while determining the enthalpy.

TABLE 4.3 Nomenclature for Program 4.3

Nomenclature	Program symbol	Description and units
h_1	A	Initial steam enthalpy (Btu/lb)
h_2	B	Final steam enthalpy (Btu/lb)
h_f	S	Water enthalpy (Btu/lb)
P	P	Steam pressure (psia)
R	R	Ratio W_f/W_s (%)
t_s	F	Saturation temperature (°F)
t_1	I	Initial steam temperature (°F)
t_2	F	Final steam temperature (°F)
W_f	—	Water for injection (lb/hr)
W_s	—	Final steam flow (lb/hr)

The enthalpy of water h_f is estimated for temperatures below 240°F as

$$h_f = t - 32 \tag{3}$$

For temperatures above 240°F,

$$h_f = 427.5 - 3.897t + 187.8 \times 10^{-4} t^2 - 30.9 \times 10^{-6} t^3 + 1.88 \times 10^{-8} t^4 \tag{4}$$

If the final steam is in saturated condition, the input of 0 for temperature will enable the program to estimate the temperature using the relation

$$t_s = 115P^{0.225} \tag{5}$$

Figure 4.3B gives the listing of the program along with the results.

Example 1

Determine the ratio of water to steam required to desuperheat steam at 1500 psia, 900°F to 820°F using feed water at 400°F.

Solution

Key in the program. In the RUN mode, the display asks for the following data: steam pressure (1500), spray water temperature (400), initial temperature (900), and final temperature (820). Once these are fed in, the program computes the enthalpies and the ratio W_f/W_s. It is seen that the initial and final enthalpies are 1431 and 1379 Btu/lb and ratio W_f/W_s = 4.92%, from the results in Fig. 4.3B.

Example 2

At a steam pressure of 300 psia, determine the water required to desuperheat steam at 600°F to saturated conditions. Feed water is at 200°F.

Solution

Input the data as before and the final steam temperature as 0. The program computes the saturation temperature as 415°F. W_f/W_s = 9.9%.

```
10 PRINT"SPRAY WATER CALCULATIONS"
20 INPUT"STM PRESS-PSIA,SPRAY WATER TEMP,INITIAL STEAM TEMP,FINAL STEAM TEMP(IF
SATURATED INPUT 0 AND PROGRAM COMPUTES THE VALUE)=";P,TW,I,F
30 IF F=0 THEN F=115*P^.225
40 T=F:GOSUB 500
50 B=Z:T=I:GOSUB 500
60 A=Z:IF TW<240 THEN S=TW-32:GOTO 80
70 Q=.01*TW:S=427.5-389.7*Q+187.8*Q*Q-30.9*Q^3+1.88*Q^4
80 R=100*(A-B)/(A-S)
90 PRINT" "
100 PRINT"RESULTS"
110 PRINT" "
120 PRINT"STM PRESS-PSIA=";P;"INITIAL STM TEMP=";I;"FINAL STM TEMP=";F
130 PRINT" "
140 PRINT"SPRAY WATER TEMP=";TW;"INITIAL ENTH -BTU/LB=";A;"FINAL ENTH-BTU/LB=";B

150 PRINT" "
160 PRINT"SPRAY TO TOTAL STM RATIO%=";R
165 PRINT" "
170 END
500 T=273.1+(T-32)/1.8:P=P/14.696
510 K=80870!/T/T:L=10^K*(-2641.62/T):M=1.89+L:N=M*P*P/T/T
520 O=2+(372420!/T/T):Q=O*L:R=1.89+Q:U=(.21828*T-126970!/T):V=2*U*R-(M/T)*126460
!
530 W=82.54-162460!/T:Y=2*W*R-(M/T)*162460!
550 Z=775.6+.63596*T+1.62467E-04*T*T+20.5697*LOG(T)
560 Z=Z+.043557*(R*P+.5*N*(Y+M*(W+V*N)))
590 P=14.696*P:T=(T-273.15)*1.8+32
600 RETURN

STM PRESS-PSIA= 1500 INITIAL STM TEMP= 900 FINAL STM TEMP= 820

SPRAY WATER TEMP= 400 INITIAL ENTH -BTU/LB= 1431.464 FINAL ENTH-BTU/LB=
 1379.542

SPRAY TO TOTAL STM RATIO%= 4.924883



SPRAY WATER CALCULATIONS
STM PRESS-PSIA,SPRAY WATER TEMP,INITIAL STEAM TEMP,FINAL STEAM TEMP(IF SATURATED
 INPUT 0 AND PROGRAM COMPUTES THE VALUE)=? 300,200,600,0

RESULTS

STM PRESS-PSIA= 300 INITIAL STM TEMP= 600 FINAL STM TEMP= 415.0019

SPRAY WATER TEMP= 200 INITIAL ENTH -BTU/LB= 1316.414 FINAL ENTH-BTU/LB=
 1202.488

SPRAY TO TOTAL STM RATIO%= 9.920306
```

FIG. 4.3B Listing of program 4.3, with results.

Reference

V. Ganapathy, *Applied Heat Transfer*, Pennwell Books, Tulsa, Oklahoma, 1982, p. 114.

PROGRAM 4.4 FLASH STEAM CALCULATION FROM BOILER BLOWDOWN

Input

Saturated water flow (blowdown)
Inlet pressure
Final pressure

Output

Enthalpy of blowdown
Enthalpy of flash steam
Enthalpy of flash liquid
Temperature of flash steam
Temperature of blowdown water
Flash steam produced

Remarks

ASME 1967 formulations are used to determine steam properties.

Flash steam recovery from boiler blowdown is often resorted to in process plants from energy conservation considerations. With this program, one can estimate the flash steam produced from saturated hot water or boiler blowdown given the blowdown quantity, initial pressure, and flash steam pressure. ASME 1967 formulations are used in the determination of the steam properties.

Theory

If x is the quantity of flash steam produced per pound of blowdown, it can be shown from energy balance that the enthalpy of blowdown water is related to flash steam enthalpy as

$$h_b = xh_g + (1 - x)h_f \tag{1}$$

Then

$$x = \frac{h_b - h_f}{h_g - h_f} \qquad (2)$$

Hence the flash steam produced

$$V = Wx \qquad (3)$$

where W is the blowdown quantity. Table 4.4 gives the nomenclature, and Fig. 4.4 shows the listing of the program.

Example

Estimate the flash steam produced when 1500 lb/hr of boiler blowdown at 1200 psia is flashed to steam at 150 psia.

Solution

Key in the program. In the RUN mode, the display asks for the following data in the same order given here: saturated liquid flow (1500), pressure 1 (1200), and pressure 2 (150). Then, the computer goes on to calculate the various enthalpies and prints out the results. As seen from Fig. 4.7, the flash steam produced is 424 lb/hr, flash steam temperature is 355° F, blowdown water temperature is 566° F. The enthalpies are also printed out.

TABLE 4.4 Nomenclature for Program 4.4

Nomenclature	Program symbol	Description and units
h_f	C	Enthalpy of flash liquid (Btu/lb)
h_b	A	Enthalpy of blowdown (Btu/lb)
h_g	B	Enthalpy of flash steam (Btu/lb)
V	V	Flash steam flow (lb/hr)
W	W	Blowdown quantity (lb/hr)
—	Z	Temperature of blowdown (°F)
—	G	Temperature of flash steam (°F)
x	X	Fraction of flash steam produced

```
10 PRINT"FLASH STEAM CALCULATIONS"
15 PRINT" "
20 INPUT"BLOW DOWN FLOW,BLOW DOWN PRESS-PSIA,FLASH PRESS-PSIA=";W,P,Q
30 G=115*P^.225:Z=G:GOSUB 700
40 A=N:D=P:G=115*Q^.225:GOSUB 700
50 C=N:P=Q:T=G:GOSUB 500
60 B=H:X=(A-C)/(B-C):V=W*X
70 PRINT" "
80 PRINT"RESULTS"
90 PRINT" "
100 PRINT"BLOW DOWN-LB/H=";W;"FLASH STEAM-LB/H=";V;"BLOW DOWN TEMP=";Z
110 PRINT" "
120 PRINT"BLOW DOWN PRES-PSIA=";D;"FLASH PRESS-PSIA=";P;"FLASH TEMP=";G
130 PRINT" "
140 PRINT"ENTH OF BLOW DOWN-BTU/LB=";A
150 PRINT" "
160 PRINT"ENTH FLASH LIQ=-BTU/LB=";C;"FLASH STEAM-BTU/LB=";B
170 PRINT" "
180 END
500 T=273.1+(T-32)/1.8:P=P/14.696
510 K=80870!/T/T:L=10^K*(-2641.62/T):M=1.89+L:N=M*P*P/T/T
520 O=2+(372420!/T/T):Q=O*L:R=1.89+Q:U=(.21828*T-126970!/T):V=2*U*R-(M/T)*126460
!
530 W1=82.54-162460!/T:Y=2*W1*R-(M/T)*162460!
540 H=775.6+.63596*T+1.62467E-04*T*T+20.5697*LOG(T)
550 H=H+.043557*(R*P+.5*N*(Y+M*(W1+V*N)))
560 P=14.696*P:T=(T-273.15)*1.8+32
570 RETURN
700 IF T<360 THEN 740
710 N=-904.11706#+10.673802#*G-.042753836#*G*G+9.41244*10^(-5)*G^3-1.0315357#*10
^(-7)*G^4
720 N=N+4.569246*10^(-11)*G^5
730 GOTO 760
740 N=-32.179105#+1.0088084#*G-1.1516996#*10^(-4)*G*G+4.855383*10^(-7)*G^3
750 N=N-7.3618778#*10^(-10)*G^4+9.6350315#*10^(-13)*G^5
760 RETURN

FLASH STEAM CALCULATIONS

BLOW DOWN FLOW,BLOW DOWN PRESS-PSIA,FLASH PRESS-PSIA=? 1500,1200,150

RESULTS

BLOW DOWN-LB/H= 1500 FLASH STEAM-LB/H= 423.5031 BLOW DOWN TEMP= 566.9093

BLOW DOWN PRES-PSIA= 1200 FLASH PRESS-PSIA= 150 FLASH TEMP= 355.0736

ENTH OF BLOW DOWN-BTU/LB= 571.5516

ENTH FLASH LIQ=-BTU/LB= 326.9738 FLASH STEAM-BTU/LB= 1193.241
```

FIG. 4.4 Listing of program 4.4, with results.

PROGRAM 4.5 STEAM FLOW IN BOILER BLOWOFF LINES

Input

Steam pressure
Pipe inner diameter
Total resistance of piping

Output

Specific volume of steam
Steam flow during sonic conditions

During start-up of boiler plants, steam blowing of lines is performed to clean the piping of mill scales, debris, and other matter. The general procedure is to operate the boiler at low pressure and allow the steam to escape to the atmosphere. Under these conditions, sonic flow is usually achieved at the pipe exit, and the usual formula that relates flow with pressure drop is not applicable. With the help of this program, one can determine the steam flow rather easily. Too high a steam flow results in pipe erosion; too low a flow may not clean the pipes adequately.

Theory

The basic equation that relates the steam flow with pressure drop for a compressible fluid is*

$$W = 1891d^2 Y \left(\frac{\Delta P}{Kv}\right)^{0.5}$$

where

d = pipe inner diameter (in.)
K = resistance coefficient of entire piping, including valves, fittings = $12fL_e/d$
L_e = total equivalent length (ft)
ΔP = maximum pressure drop resulting in sonic flow = $(\Delta P/P_1)P_1$. The factor $(\Delta P/P_1)$ is a function of K*
P_1 = drum pressure (psia)

*Crane Technical paper 410, Flow of fluids, 1981, p. 413.

```
10 PRINT"FLOW DURING STEAM LINE BLOWING OPERATIONS"
15 PRINT" "
20 INPUT"STM PRESS-PSIA,PIPE ID,K FACTOR(F*L/D)=";P,D,K
25 IF K=1 THEN K=K+.1
30 T=115*P^.225:GOSUB 500
40 X=P*(.1685*LOG(LOG(K)/2.3025)+.776)
45 PRINT" "
50 L=P-X:IF L<=15 THEN PRINT" CONVENTIONAL FORMULA FOR FLOW USED:(STM PR-15) FOR
 DIFFERENTIAL PRESSURE":GOTO 75
60 Y=.703+.0562*LOG(LOG(K)/2.3025)
70 W=1891*Y*D*D*(X/K/V)^.5:GOTO 80
75 W=1891*D*D*((P-15)/K/V)^.5
80 PRINT" "
90 PRINT"STM PRESS-PSIA=";P;"SAT TEMP=";T;"SP VOL-CU FT/LB=";V
100 PRINT" "
110 PRINT"PIPE ID=";D;"K FACTOR=";K;"STM FLOW-LB/H=";W
120 PRINT" "
130 END
500 T=.01*T:IF T>4.6 GOTO 540
510 IF T>3 GOTO 530
520 V=80.82*T^4-767.06*T^3+2726.21*T*T-4337.7*T+2647.9:GOTO 550
530 V=.529*T^4-9.208*T^3+60.986*T*T-183.361*T+213.456:GOTO 550
540 V=.02097*T^4-.542*T^3+5.319*T*T-23.676*T+40.709
550 T=100*T:RETURN

FLOW DURING STEAM LINE BLOWING OPERATIONS

STM PRESS-PSIA,PIPE ID,K FACTOR(F*L/D)=? 170,2.067,11.84


STM PRESS-PSIA= 170 SAT TEMP= 365.2152 SP VOL-CU FT/LB= 2.800354

PIPE ID= 2.067 K FACTOR= 11.84 STM FLOW-LB/H= 11480.66
```

FIG. 4.5 Listing of program 4.5 (steam blowing), with results.

v = steam specific volume at P_1, (ft^3/lb)
f = Darcy friction factor
Y = expansion factor, a function of K*

A check is usually made to see if the downstream pressure obtained from the maximum pressure drop criterion is greater than the atmospheric pressure. If it is greater, then sonic flow results and the procedure is used to determine steam flow. If not, then the computer indicates that the usual equations of flow are applicable and program 2.1 is used. Figure 4.5 shows the listing of the program.

*Crane Technical paper 410, Flow of fluids, 1981, p. 4.13.

Example

In a boiler plant, steam blowing is done at 170 psia. If the pipe inner diameter is 2.067 in. and K = 11.84, determine the flow.

Solution

Key in the program. The screen asks for the line size, steam pressure, and K factor. Then, the results are printed out. It is seen that the steam flow is 11480 lb/hr.

5

Performance of Heat Transfer Equipment

PROGRAM 5.1 PERFORMANCE OF FIRE TUBE WASTE HEAT BOILERS

Input

Tube OD
Tube ID
Tube length
Number of tubes
Gas temperature in
Steam pressure
Feed water temperature
Gas flow

Output

Steam generation
Gas temperature at boiler exit
Gas pressure drop
Duty (MM Btu/hr)

Remarks

Saturation temperature and enthalpy of feed water and steam are estimated by the program.

Fire tube waste heat boilers (Fig. 5.1A) are widely used in process plants for recovering energy from waste gas streams. It is desirable to be able to predict their performance under different conditions of gas flow, gas inlet temperature, and steam pressure. Also, comparison may be made between measured and predicted performance to see whether the equipment is operating well.

The program calculates the exit gas temperature, duty, steam generation, and gas pressure drop given the gas flow, gas inlet temperature, tube configuration, steam pressure, and feed water temperature. Good estimates are obtained with flue gases from the combustion of fossil fuels, turbine exhaust, and effluents from chemical plants at atmospheric pressure.

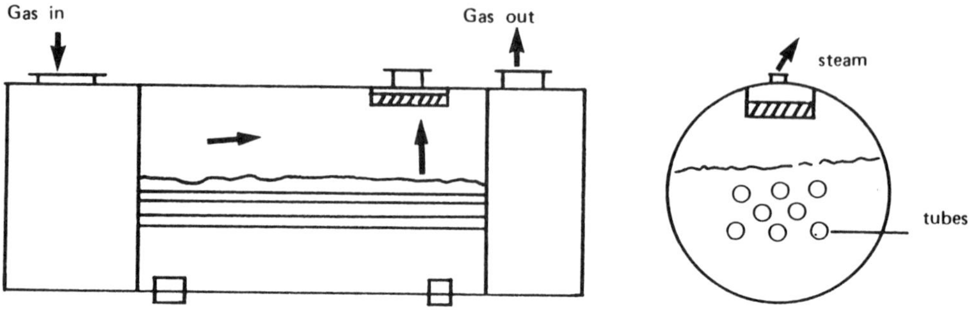

FIG. 5.1A Fire tube waste heat boiler.

Theory

The energy transferred Q is given by the equation*†:

$$Q = W_g C_p(t_1 - t_2) = \frac{UA(t_1 - t_2)}{\ln\,[(t_1 - t_s)/(t_2 - t_s)]} = W_s(h_g - h_w) \tag{1}$$

The overall heat transfer coefficient U is given by

$$\frac{1}{U} = \frac{1d}{h_i d_i} + ff_i \frac{d}{d_i} + \frac{d}{24K_m} \ln \frac{d}{d_i} + ff_o + \frac{1}{h_o} \tag{2}$$

The gas-side heat transfer coefficient h_i controls U and may be obtained as

$$h_i = 2.44\left(\frac{W_g}{n}\right)^{0.8} \frac{0.163 + 0.000021(t_1 + t_2)}{d_i^{1.8}} \tag{3}$$

The effect of outside heat transfer coefficient h_o is not significant; using nominal fouling factors, the equations may be simplified into

$$\frac{1}{U} = \frac{d}{h_i d_i} + d\,\frac{\ln(d/d_i)}{600} + 0.003 \tag{4}$$

---

*V. Ganapathy, Size or check waste heat boilers quickly, *Hydrocarbon Processing*, September 1984, p. 169.

†V. Ganapathy, *Applied Heat Transfer*, Pennwell Books, Tulsa, Oklahoma, 1982, pp. 160-189.

The external surface area

$$A = 3.14 \frac{dL}{12} \tag{5}$$

The enthalpy absorbed by steam may be estimated as

$$h_g - h_w = 1031 + 1.13t_s - 0.0045t_s^2 + 10.8 \times 10^{-6} t_s^3 - 959 \times 10^{-8} t_s^4 - t_w + 32 \tag{6}$$

where saturation temperature

$$t_s = 115P^{0.225} \tag{7}$$

The flue gas specific heat

$$C_p = 0.2465 + 0.00002125(t_1 + t_2) \tag{8}$$

The duty Q is arrived at through a trial-and-error procedure, which will be discussed later.

The gas pressure drop through the tubes is obtained from

$$\Delta P = 9.3 \times 10^{-5} f \left(\frac{W_g}{n}\right)^2 \frac{L + 5d_i}{\rho_g d_i^5} \tag{9}$$

where friction factor f is calculated as

$$f = 0.0055 \left\{1 + \left[\left(\frac{36}{d_i}\right) + 1\right]^{0.33}\right\} \tag{10}$$

The gas density

$$\rho_g = \frac{40}{460 + t_m} \tag{11}$$

where t_m is the average gas temperature. Table 5.1 shows the nomenclature.

Use of Program

The program adopts a trial-and-error routine to arrive at the performance. The steps are

1. Input the tube configuration, steam pressure, gas flow, and feed water temperature.
2. Assume a value of t_2; the first trial value is $t_s + 1$.

TABLE 5.1 Nomenclature for Program 5.1

Nomenclature	Program symbol	Description and units
d_i	I	Tube inner diameter (in.)
d	D	Tube outer diameter (in.)
f	F	Friction factor
h_i	H	Gas heat transfer coefficient (Btu/ft^2 hr °F)
L	L	Tube length (ft)
n	N	Number of tubes in each pass
P	P	Steam pressure (psia)
Q	Q	Duty (MM Btu/hr)
t_1	A	Gas temperature in (°F)
t_2	B	Gas temperature out (°F)
t_s	S	Saturation temperature (°F)
t_w	C	Feed water temperature (°F)
U	U	Overall heat transfer coefficient (Btu/ft^2 hr °F)
W_g	W	Gas flow (lb/hr)
W_s	O	Steam flow (lb/hr)
ρ_g	Z	Gas density (lb/ft^3)
ΔP	G	Gas pressure drop (in. WC)

3. Calculate C_p and Q from equation (1a).
4. Calculate h_i, U, A, and t_s.
5. Calculate Q from equation (1b).
6. If Q calculated from steps 3 and 5 are equal, the program goes on to calculate W_s and gas pressure drop. If not, the assumed t_2 is corrected. The logic used helps arrive at the correct value of t_2 in a few trials. The second trial value of $t_2 = (t_s - 1) \pm 0.5(t_2 - t_1)$, depending on whether the assumed value was low or high. Then, steps 3-6 are repeated.
7. The results are printed out.

```
10 PRINT"FIRE TUBE WASTE HEAT BOILER"
20 INPUT"TUBE OD,TUBE ID,LENGTH,NO OF TUBES=";D,I,L,N
30 INPUT"GAS TEMP IN,STM PRESS-PSIA,FEED WATER IN,GAS FLOW=";A,P,C,W
40 S=115*P^.225:B=S+1:V=A-B
45 Q=W*(A-B)*(.2465+2.125E-05*(A+B))
50 H=2.44*(W/N)^.8*(.163+.000021*(A+B))/I^1.8
60 U=1/((D/I/H)+(D/600)*LOG(D/I)+.003):R=U*(3.14*D/12)*N*L*(A-B)/LOG((A-S)/(B-S)
)
70 IF ABS((Q-R)/R)<.02 THEN 100
80 V=.5*V:B=B+SGN(Q-R)*V:IF B<S THEN B=S+1
90 GOTO 45
100 M=.01*S:O=Q/(1031+113.3*M-45.3*M*M+10.8*M^3-.959*M^4-C+32)
110 F=.0055*(1+((36/I)+1)^.33):Z=40/(460+.5*(A+B))
120 G=9.3*10^(-5)*F*(W/N)*(W/N)*(L+5*I)/Z/I^5:Q=Q*10^(-6)
125 PRINT" "
130 PRINT"RESULTS"
140 PRINT" "
150 PRINT"GAS FLOW=";W;"GAS TEMP IN=";A;"TEMP OUT=";B;"STM PRESS-PSIA=";P
160 PRINT" "
170 PRINT"FEED WATER IN=";C;"STEAM FLOW=";O;"DUTY-MMBTU/H=";Q;"STM TEMP=";S
180 PRINT" "
190 PRINT"TUBE OD=";D;"TUBE ID=";I;"LENGTH=";L;"NUMBER=";N
200 PRINT" "
210 PRINT"OV HT TR COEFF-BTU/FT2HF=";U;"GAS PR DROP-IN WC=";G
220 END
```

```
FIRE TUBE WASTE HEAT BOILER
TUBE OD,TUBE ID,LENGTH,NO OF TUBES=? 2,1.726,12.9,440
GAS TEMP IN,STM PRESS-PSIA,FEED WATER IN,GAS FLOW=? 1750,400,220,65000

RESULTS

GAS FLOW= 65000 GAS TEMP IN= 1750 TEMP OUT= 770.3146 STM PRESS-PSIA= 400

FEED WATER IN= 220 STEAM FLOW= 18633.02 DUTY-MMBTU/H= 19.10748 STM TEMP=
 442.7528

TUBE OD= 2 TUBE ID= 1.726 LENGTH= 12.9 NUMBER= 440

OV HT TR COEFF-BTU/FT2HF= 8.969839 GAS PR DROP-IN WC= 2.541873
```

FIG. 5.1B Listing of program 5.1, with results.

Example

A fire tube waste heat boiler is used for recovering energy from an incinerator. The data are

1. Tube OD = 2.0 in.
2. Tube ID = 1.716 in.
3. Tube length = 12.9 ft

4. Number of tubes = 440
5. Gas temperature in = 1750°F
6. Steam pressure = 400 psia (385 psig)
7. Feed water temperature = 220°F
8. Gas flow = 65,000 lb/hr

Determine the exit gas temperature, duty, steam generation, and gas pressure drop that may be expected.

Solution

Key in the program, a listing of which is given in Fig. 5.1B. The screen asks for the data in the same order given here, they are fed in. Then, using the logic discussed, the computer solves the various equations and prints out the results. It is seen that the exit gas temperature is 770°F, duty is 19 MM Btu/hr, steam flow is 18,633 lb/hr, and gas pressure drop is 2.6 in. WC.

PROGRAM 5.2 PERFORMANCE OF WATER TUBE WASTE HEAT BOILERS

Input

Tube OD
Tube ID
Transverse and longitudinal pitch
Tube length
Number of tubes wide
Number of tubes deep
Partial pressures of CO_2 and H_2O
Arrangement—inline or staggered
Gas flow
Gas temperature in
Steam pressure
Feed water temperature

Output

Duty (MM Btu/hr)
Steam generation
Gas temperature at boiler exit
Overall heat transfer coefficient
Gas pressure drop

Remarks

Valid for bare tube boilers in inline or staggered arrangement. The program gives good estimates for products of the combustion of fossil fuels.

This program predicts the performance of waste heat boilers of natural or forced circulation design using bare tubes. Figure 5.2A shows the arrangement of a typical boiler. Exit gas temperature, duty, steam generation, and gas pressure drop are predicted by the program, given the gas flow, inlet gas temperature, and tube geometry. Tube banks could be inline or in a staggered arrangement. Good estimates are obtained for flue gases from products of the combustion of fossil fuels. Convective and nonluminous heat transfer coefficients are considered.

Theory

The energy transferred Q is given by*:

$$Q = W_g C_p (t_1 - t_2) = \frac{UA(t_1 - t_2)}{\ln [(t_1 - t_s)/(t_2 - t_s)]} = W_s(h_g - h_w) \tag{1}$$

The overall heat transfer coefficient U is given by

$$\frac{1}{U} = \frac{1}{h_c + h_n} + \frac{ff_i d}{d_i} + ff_o + d\,\frac{\ln (d/d_i)}{24K_m} + \frac{d}{h_i d_i} \tag{2}$$

This equation may be simplifed as follows, bearing in mind that the tube outside resistance to heat transfer and fouling factors is small, about 0.003.

$$\frac{1}{U} = \frac{1}{h_c + h_n} + d\,\frac{\ln (d/d_i)}{600} + 0.003 \tag{3}$$

The convective heat transfer coefficient h_c is obtained from*

$$h_c = \frac{0.9G^{0.6}(k^{0.67}C_p{}^{0.33}/\mu^{0.27})}{d^{0.4}} = \frac{0.9G^{0.6}F}{d^{0.4}} \tag{4}$$

*V. Ganapathy, *Applied Heat Transfer*, Pennwell Books, Tulsa, Oklahoma, 1982, p. 190.

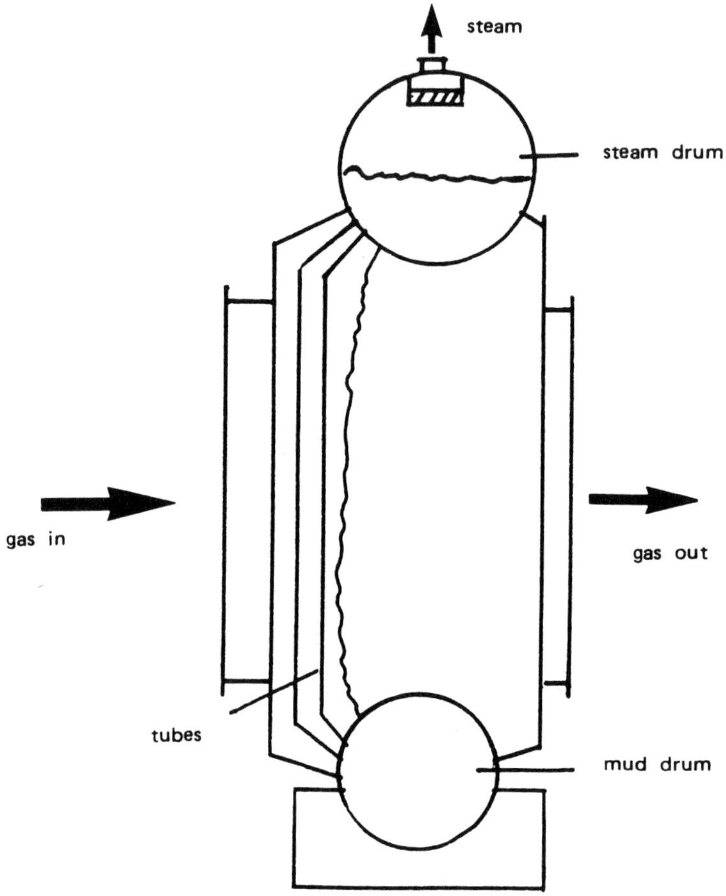

FIG. 5.2A Water tube waste heat boiler.

The term involving gas properties is evaluated at the gas film temperature t as

$$F = 0.094 + 0.00004(t_f - 200) \tag{5}$$

This is valid for common flue gases at atmospheric pressure.

The film temperature t_f is approximated as $t_f = 0.5(t_g + t_s)$. The gas mass velocity

$$G = \frac{12W_g}{N_w L(s_t - d)} \tag{6}$$

The nonluminous heat transfer coefficient h_n is obtained as discussed in program 3.6. The average gas temperature is $0.5(t_1 + t_2)$ and surface temperature is taken as $t_s + 40$. Surface area

$$A = \frac{3.14 N_w d L N_d}{12} \tag{7}$$

Partial pressures of CO_2 and H_2O needed for estimating h_n may be obtained from Table 3.7 (program 3.6) if not known. Gas specific heat is estimated as

$$C_p = 0.2465 + 0.00002125(t_1 + t_2) \tag{8}$$

Enthalpy absorbed by steam $(h_g - h_w)$ and saturation temperature t_s are obtained from equations discussed in program 5.1.

The gas pressure drop over bare tube bundles is obtained as discussed in program 2.8. Viscosity of flue gas is obtained from

$$\mu = 0.0619 + 0.000021(t_1 + t_2) \tag{9}$$

Gas density

$$\rho_g = \frac{40}{460 + 0.5(t_1 + t_2)} \tag{10}$$

Use of Program

The program adopts a trial-and-error logic to arrive at the solution. The steps are

1. The tube configuration, gas inlet conditions, steam pressure, and feed water temperature are input.
2. t_2 is assumed. First trial value is $t_s + 1$.
3. t_f, t_s, h_c, h_n, and U are calculated using the equations discussed.
4. Q is calculated using equation (1a).
5. Q is estimated from equation (1b).
6. If Q from steps 4 and 5 are equal, the assumed t_2 is correct. The program goes to step 6. Otherwise, t_2 is corrected. The logic used is such that the result is arrived at in a few steps. Then steps 3-6 are repeated.

TABLE 5.2 Nomenclature for Program 5.2

Nomenclature	Program symbol	Description and units
d	D	Tube outer diameter (in.)
d_i	I	Tube inner diameter (in.)
f	F	Friction factor
G	G	Gas mass velocity (lb/ft^2 hr)
h_c	H	Convective heat transfer coefficient (Btu/ft^2 hr °F)
h_g	—	Enthalpy of saturated steam (Btu/lb)
h_n	A28	Nonluminous heat transfer coefficient (Btu/ft^2 hr °F)
h_w	—	Feed water enthalpy (Btu/lb)
L	L	Tube length (ft)
N_d	K	Number of tubes deep
N_w	N	Number wide
P	P	Steam pressure (psia)
Q	Q	Duty (MM Btu/hr)
s_t	X	Transverse pitch (in.)
s_l	Y	Longitudinal pitch (in.)
t_1	A	Gas inlet temperature (°F)
t_2	B	Gas exit temperature (°F)
t_f	—	Gas film temperature (°F)
t_g	—	Average gas temperature (°F)
t_w	C	Feed water temperature (°F)
U	U	Overall heat transfer coefficient (Btu/ft^2 hr °F)
W_g	W	Gas flow (lb/hr)
W_s	O	Steam flow (lb/hr)
ρ_g	—	Gas density (lb/ft^3)
μ	—	Gas viscosity (lb/ft hr)
ΔP_g	A30	Gas pressure drop (in. WC)

7. The results are printed out. Alternatively, the desired result may be obtained by pressing the appropriate key. See Table 5.2 for nomenclature.

Example

A water tube waste heat boiler with bare tubes has the data

1. Tube OD = 2.0 in.
2. Tube ID = 1.686 in.
3. Transverse pitch = 4.0 in.
4. Longitudinal pitch = 4.0 in.
5. Tube length = 13.5 ft
6. Number of tubes wide = 18
7. Number deep = 16
8. Partial pressure of CO_2 = 0.15
9. Partial pressure of H_2O = 0.12
10. Arrangement (inline = 1, staggered = 0) = 1
11. Gas flow = 110,000 lb/hr
12. Gas temperature in = 1520°F
13. Steam pressure = 400 psia
14. Feed water temperature = 225°F

Predict the exit gas temperature, duty, steam generation, and gas pressure drop.

Solution

Key in the program, listing of which is given in Fig. 5.2B. In the RUN mode, the display asks for the data in the same order given here; they are fed in. The computer goes on to solve for t_2 using the trial-and-error logic discussed and prints out the results. It is seen that exit gas temperature = 998°F, steam generation is 16,865 lb/hr, gas pressure drop is 0.269 in. WC, and the duty is 17.2 mm Btu/hr. Note that blowdown was not considered in evaluating the steam flow.

Caution: If partial pressures of CO_2 and H_2O are zero, use a value of 0.0001 to avoid execution errors.

```
10 PRINT"PERFORMANCE OF WATER TUBE WASTE HEAT BOILERS"
20 INPUT"TUBE OD,TUBE ID,TR PITCH,LONG PITCH,TUBE LENGTH=";D,I,X,Y,L
30 INPUT"NO OF TUBES WIDE,NO DEEP,PCO2,PH2O(IF PCO2,PH2O ARE NIL USE A VALE OF 0
.0001 TO EXECUTE PROGRAM";N,K,M,T
40 INPUT"ARRANGEMENT-INPUT 0 FOR STAGGERED OR 1 FOR INLINE=";J
50 INPUT"GAS FLOW,GAS TEMP IN,STM PRESS-PSIA,FEED WATER IN=";W,A,P,C
60 S=115*P^.225:B=S+1:V=A-B
70 Q=W*(A-B)*(.2465+2.125E-05*(A+B)):F=(.5*(A+B)+S)*.5:G=12*W/(N*L*(X-D))
80 H=.9*G^.6*(9.399999E-02+.00004*(F-200))/D^.4:E=1.08*(X*Y-.785*D*D)/D/39.36
90 A27=((.5*(A+B)-32)/1.8)+273:A28=(.8+1.6*T)*(1-.00038*A27)*(M+T)/((M+T)*E)^.5
100 A28=(1-EXP(-A28*E))*10^(-8)*((.5*(A+B)+460)^4-(S+500)^4)/(.5*(A+B)-S-40)
110 A28=.1401*A28:U=1/((1/(H+A28))+.003+D*LOG(D/I)/600)
120 R=U*(3.14*D/12)*N*L*K*(A-B)/LOG((A-S)/(B-S))
130 IF ABS((Q-R)/R)<.02 THEN 170
140 V=.5*V:B=B+SGN(Q-R)*V:IF B<S THEN 160
150 GOTO 70
160 B=S+1
170 Z=.01*S:O=Q/(1031+113.3*Z-45.32*Z*Z+10.8*Z^3-.959*Z^4-C+32)
180 R=G*D/12/(.0619+.000042*(.5*(A+B)-500))
190 IF J=0 THEN F=R^(-.16)*(.25+(.1175/((X/D)-1)^1.08)):GOTO 210
200 F=R^(-.15)*(.044+((.08*Y/D)/((X/D)-1)^(.43+1.13*D/Y)))
210 A30=9.3*10^(-10)*F*G*G*K/(40/(460+.5*(A+B))):Q=Q/10^6
215 PRINT" "
220 PRINT"RESULTS"
230 PRINT" "
240 PRINT"GAS FLOW=";W;"GAS TEMP IN=";A;"GAS OUT=";B;"DUTY-MMBTU/H=";Q
250 PRINT" "
260 PRINT"SAT TEMP=";S;"FEED WATER TEMP=";C;"STEAM FLOW=";O
270 PRINT" "
280 PRINT"TUBE OD=";D;"TUBE ID=";I;"TR PITCH=";X;"LONG PITCH=";Y
290 PRINT" "
300 PRINT"TUBE LENGTH=";L;"NO OF TUBES WIDE=";N;"NO DEEP=";K
310 PRINT" "
320 PRINT"OV HT TR COEFF-BTU/FT2HF=";U;"GAS PR DROP-IN WC=";A30
330 PRINT" "
340 IF J=1 THEN PRINT"INLINE ARRANGEMENT":GOTO 360
350 PRINT"STAGGERED ARRANGEMENT"
360 END
```

FIG. 5.2B Listing of program 5.2, with results.

PROGRAM 5.3 PERFORMANCE OF ECONOMIZERS

Input

Gas flow
Gas temperature in
Water flow
Tube OD
Water temperature in

```
PERFORMANCE OF WATER TUBE WASTE HEAT BOILERS
TUBE OD,TUBE ID,TR PITCH,LONG PITCH,TUBE LENGTH=? 2,1.686,4,4,13.5
NO OF TUBES WIDE,NO DEEP,PCO2,PH2O(IF PCO2,PH2O ARE NIL USE A VALE OF Ø.ØØØ1 TO
EXECUTE PROGRAM? 18,16,.15,.12
ARRANGEMENT-INPUT Ø FOR STAGGERED OR 1 FOR INLINE=? 1
GAS FLOW,GAS TEMP IN,STM PRESS-PSIA,FEED WATER IN=? 11ØØØØ,152Ø,4ØØ,225

RESULTS

GAS FLOW= 11ØØØØ GAS TEMP IN= 152Ø GAS OUT= 998.6926 DUTY-MMBTU/H= 17.2Ø442

SAT TEMP= 442.7528 FEED WATER TEMP= 225 STEAM FLOW= 16865.9

TUBE OD= 2 TUBE ID= 1.686 TR PITCH= 4 LONG PITCH= 4

TUBE LENGTH= 13.5 NO OF TUBES WIDE= 18 NO DEEP= 16

OV HT TR COEFF-BTU/FT2HF= 1Ø.82169 GAS PR DROP-IN WC= .2696926

INLINE ARRANGEMENT
```

FIG. 5.2B (Continued)

Tube ID
Fin density
Fin height
Fin thickness
Transverse pitch
Longitudinal pitch
Length
Number of tubes wide
Parallel passes
Tubes deep

Output

Duty (MM Btu/hr)
Water temperature leaving
Gas temperature leaving
Water pressure drop
Gas pressure drop

Remarks

This program valid for solid finned tubes in a staggered arrangement.

Economizers are widely used in boiler plants to recover energy from exhaust gases. With the program discussed here, one can predict the performance of spiral finned economizers in a staggered arrangement. The program evaluates the energy absorbed by a given economizer, the exit gas and water temperatures, and water and gas pressure drops. A trial-and-error procedure is used to arrive at the solution. The program is adequate for the products of combustion of fossil fuels. Moderate fouling is assumed.

Theory

Energy absorbed and transferred may be obtained from

$$Q = W_g c_{pg}(T_1 - T_2) = wc_{pw}(t_2 - t_1) = US\,\Delta T \tag{1}$$

$$\Delta T = \frac{(T_1 - t_2) - (T_2 - t_1)}{\ln [(T_1 - t_2)/(T_2 - t_1)]} \tag{2}$$

$$\frac{1}{U} = 1/[(1/h_g) + 12A_t/[(ff_i + h_i)d_i] + (d/K)\ln(d/d_i)A_t/ [(d + d_i)] + ff_0] \tag{3}$$

The gas-side heat transfer coefficient h_g is obtained from the equation of Robinson and Briggs*:

$$h_g = 0.295G^{0.681}k_g^{0.67}c_{pg}(1/n - b)^{0.313}/ (d^{0.319}\mu_g^{0.351}h^{0.2}b^{0.313}) \tag{4}$$

$$G = \frac{W_g}{N_w L(s_t - A_0)} \tag{5}$$

$$A_0 = \frac{d}{12} + \frac{nbh}{6} \tag{6}$$

Fin efficiency is obtained from

$$\phi = \frac{1}{1 + m^2h^2[(d + 2h)/d]^{0.5}/3} \tag{7}$$

*V. Ganapathy, Charts simplify spiral finned tube calculations, *Chemical Engineering*, April 25, 1977.

where $m = (24h_g/Kb)^{0.5}$, K being the thermal conductivity of fin material. The fin effectiveness

$$\eta = 1 - (1 - \phi)\frac{A_f}{A_t} \tag{8}$$

The fin and total surface area per unit length may be obtained from

$$A_f = \frac{\pi n(4dh + 4h^2 + 2bd + 4bh)}{24} \tag{9}$$

$$A_t = A_f + (1 - nb)\frac{\pi d}{12} \tag{10}$$

The factor that accounts for the gas properties in equation (4) has been estimated as follows for common flue gases, as well as air, to a reasonable accuracy:

$$F = 0.125 + 0.00004t = \frac{k_g{}^{0.67} c_{pg}{}^{0.33}}{\mu_g{}^{0.351}} \tag{11}$$

Tubeside heat transfer coefficient h_i is obtained from*

$$h_i = 2.44\left(\frac{w}{p}\right)^{0.8}\frac{F(t)}{d_i{}^{1.8}} \tag{12}$$

$$F(t) = k^{0.6}\left(\frac{c_p}{\mu}\right)^{0.4} = 10^{(-1.318+0.493 \log t)} \quad \text{for water} \tag{13}$$

The gas pressure drop ΔP_g is calculated from the equation of Briggs and Young:

$$\Delta P_g = \frac{1.58 \times 10^{-8} G^{1.684} d^{0.611} N_d (460 + T)\mu^{0.316}}{s_t{}^{0.412} s_1{}^{0.515} MW} \tag{14}$$

This equation may be simplified for air and common flue gases as

$$\Delta P_g = \frac{51 \times 10^{-11} G^{1.684} d^{0.611} (136 + 0.528T) N_d}{s_t{}^{0.412} s_1{}^{0.515}} \tag{15}$$

*V. Ganapathy, *Applied Heat Transfer*, Pennwell Books, Tulsa, Oklahoma, 1982.

```
10 CLS:KEY OFF:PRINT"ECONOMISER PERFORMANCE-FINNED TUBE BUNDLE-STAGGERED ARRGT O
NLY"
15 PRINT" "
20 INPUT"GAS FLOW,GAS INLET TEMP,WATER FLOW,WATER INLET TEMP=";W,R,V,M
25 PRINT" "
30 INPUT"TUBE OD,ID,FINS/IN,FIN HT,FIN THK,TR PITCH,LONG PITCH=";D,I,N,H,B,X,A31

35 PRINT" "
40 INPUT"TUBE LENGTH,NO OF TUBES WIDE,NO OF PARALLEL PASSES,NO OF TUBES DEEP=";L
,A28,P,A30
50 A32=(D/12)+N*B*H/6:F=3.14*N*(4*D*H+4*H*H+2*B*D+4*B*H)/24
60 T=F+3.14*D*(1-N*B)/12:S=T*A28*A30*L
70 C=.5*W*(R-M)*(.2465+2.125E-05*(R+M)):J=.57*V*(R-M)
80 Q=(C+J-ABS(C-J))/2:Y=Q
90 K=M+Q/V
95 E=.5*(M+K)
100 GOSUB 500
110 A29=M+Q/V/Z
120 IF ABS(A29-K)<.5 THEN 140
130 K=.5*(K+A29):GOTO 95
140 IF ABS(K-R)<.2 THEN PRINT"CHECK SURFACE AREA/DATA":GOTO 370
150 O=R-4*Q/W
160 Z=.2465+2.125E-05*(R+O)
170 A29=R-Q/(W*Z)
180 IF ABS(A29-O)<.5 THEN 200
190 O=.5*(O+A29):GOTO 160
200 IF ABS(O-M)<.2 THEN PRINT"CHECK SURFACE AREA/DATA":GOTO 370
210 G=W/(A28*L*((X/12)-A32)):A27=.295*G^.681*((1/N)-B)^.313*(.125+.00004*(.5*(R+
O)-400))/D^.319/B^.113/H^.2
220 J=1/(1+(.33*(A27/B)*H*H*((D+2*H)/D)^.5)/144):J=1-(1-J)*F/T:A27=A27*J
230 A=2.44*(V/P)^.8*10^(-1.318+.2141*LOG(.5*(M+K)))/I^1.8
240 U=1/((1/A27)+(12*T/3.14/I)*((1/A)+.001)+.006*LOG(D/I)*T+.001)
250 Z=((R-K)-(O-M))/LOG((R-K)/(O-M)):A29=U*S*Z
260 IF ABS((A29-Q)/A29)<.02 THEN 280
270 Y=.5*Y:Q=Q+SGN(A29-Q)*Y:GOTO 90
280 A29=51*10^(-11)*G^1.684*D^.611*(136+.2641*(R+O))*A30/X^.412/A31^.515
290 FF=.0055*(1+((36/I)+1)^.33):C=3.36E-06*FF*(V/P)*(V/P)*.015*EXP(.000279*(M+K)
)/I^5
300 P=C*((A28/P)*A30*L+((A28/P)*A30-1)*2.5*I):Q=Q*10^-6
305 PRINT" "
310 PRINT"RESULTS"
320 PRINT" "
330 PRINT"GAS FLOW=";W;"WATER FLOW=";V;"DUTY-MM BTU/H=";Q
340 PRINT" "
350 PRINT"GAS TEMP IN=";R;" TEMP OUT=";O;"WATER TEMP IN=";M;"TEMP OUT=";K
360 PRINT" "
370 PRINT"TUBE OD=";D;"ID=";I;"TR PITCH=";X;"LONG PITCH=";A31;"FINS/IN=";N;"HT="
;H;"THK=";B
380 PRINT" "
390 PRINT"TUBE LENGTH-FT=";L;"NO WIDE=";A28;"NO DEEP=";A30
400 PRINT" "
410 PRINT"TOT SURF AREA=";S;"GAS PR DROP-IN WC=";A29;"WAT PR DROP=";P
420 PRINT" "
430 PRINT"OV HT TR COEF-BTU/FT2HF=";U;"FIN EFF=";J;"TUBE SIDE HTC=";A
435 PRINT" "
440 END
500 IF E<335 THEN 530
510 Z=.995*EXP(-2.93+.505*LOG (E))
520 GOTO 540
530 Z=1
540 RETURN
```

FIG. 5.3 Listing of program 5.3, with results.

```
GAS FLOW,GAS INLET TEMP,WATER FLOW,WATER INLET TEMP=? 130000,600,47000,220

TUBE OD,ID,FINS/IN,FIN HT,FIN THK,TR PITCH,LONG PITCH=? 2,1.78,3,.75,.06,4.5,3.9

TUBE LENGTH,NO OF TUBES WIDE,NO OF PARALLEL PASSES,NO OF TUBES DEEP=? 12,12,4,28

RESULTS

GAS FLOW= 130000 WATER FLOW= 47000 DUTY-MM BTU/H= 10.79701

GAS TEMP IN= 600  TEMP OUT= 286.6527 WATER TEMP IN= 220 TEMP OUT= 449.7236

TUBE OD= 2 ID= 1.78 TR PITCH= 4.5 LONG PITCH= 3.9 FINS/IN= 3 HT= .75 THK= .06

TUBE LENGTH-FT= 12 NO WIDE= 12 NO DEEP= 28

TOT SURF AREA= 15451.06 GAS PR DROP-IN WC= 3.479564 WAT PR DROP= 13.30187

OV HT TR COEF-BTU/FT2HF= 6.717965 FIN EFF= .8125133 TUBE SIDE HTC= 1316.379
```

FIG. 5.3 (Continued)

It gives a conservative estimate of gas pressure drop.

Tubeside pressure drop may be found from

$$\Delta P_w = 3.36 \times 10^{-6} f L_e \left(\frac{w}{p}\right)^2 \frac{v}{d_i^5} \tag{16}$$

The equivalent length

$$L_e = \left(\frac{N_w}{p}\right) N_d L + \left(\frac{N_w}{p}\right) N_d 2.5 d_i \tag{17}$$

$$f = 0.0055 \left\{ 1 + \left(\frac{36}{d_i} + 1\right)^{0.33} \right\} \tag{18}$$

More information on this design procedure can be found in *Applied Heat Transfer*.

Performance Procedure

In a performance calculation, the configuration of the economizer will be known, along with the gas, water flows, and the inlet temperatures of the fluids. A trial-and-error logic is used to arrive at the final solution. The following are the features of the program, a listing of which appears in Fig. 5.3.

TABLE 5.3 Nomenclature for Program 5.3

Nomenclature	Program symbol	Description and units
A_f	F	Fin area (ft^2/ft)
A_0	A32	Obstruction area (ft^2/ft)
A_t	T	Total area (ft^2/ft)
b	B	Fin thickness (in.)
c_p	Z	Specific heat (Btu/lb °F)
d, d_i	D,I	Tube OD, ID (in.)
f	—	Friction factor
G	G	Gas mass velocity (lb/ft^2 hr)
h_i	A	Water heat transfer coefficient (Btu/ft^2 hr °F)
h_g	A27	Gas heat transfer coefficient (Btu/ft^2 hr °F)
h	H	Fin height (in.)
k	—	Gas thermal conductivity (Btu/ft hr °F)
K	—	Metal thermal conductivity (Btu/ft hr °F)
L	L	Tube length (ft)
L_e		Equivalent length (ft)
N_w	A28	Number of tubes wide
N_d	A30	Number of tubes deep
n	N	Fins per inch
p	P	Number of parallel passes
ΔP_g, ΔP_w	A29, C	Gas per drop (in. WC), water drop (psi), energy transferred (Btu/hr)
Q	Q	Energy transferred (Btu/hr)
s_t, s_l	X, A31	Transverse and longitudinal pitch (in.)
S	S	Surface area (ft^2)
T_1, T_2	R, O	Gas temperature in, out (°F)
t_1, t_2	M, K	Water temperature in, out (°F)
ΔT	—	Log mean temperature difference (°F)

TABLE 5.3 (Continued)

Nomenclature	Program symbol	Description and units
U	U	Overall heat transfer coefficient (Btu/ft^2 hr °F)
v	—	Specific volume (ft^3/lb)
W_g, w	W, V	Gas and water flow (lb/hr)
μ	—	Viscosity (lb/ft hr)
ϕ	—	Fin efficiency
η	J	Fin effectiveness

1. A logic is used that arrives at the results quickly. The first trial value of Q is half the smaller of $W_g c_{pg}(T_1 - t_1)$ and $wc_{pw}(T_1 - t_1)$. Depending on whether the assumed Q is smaller or larger than the transferred energy $Q_t = US\ \Delta T$, the next trial value of Q is changed by half the value of Q.
2. In case T_2 falls below t_1 or t_2 goes beyond T_1, the program is stopped. This situation can occur if a very large surface area is input by error.
3. The results may be used to predict the off-design performance of economizers. Also, it can be used to check if fouling has occurred. For example, if the exit gas temperature or gas pressure drop is much more than that predicted, one can assume that fouling has occurred.

Example

An economizer has the following data:

Gas flow = 130,000 lb/hr
Gas inlet = 600°F
Water flow = 47,000 lb/hr
Water inlet = 220°F
Tube OD = 2.0 in.
Tube ID = 1.78 in.
Fins per in. = 3

Fin height = 0.75 in.
Fin thickness = 0.06 in.
Transverse pitch = 4.5 in.
Long pitch = 3.9 in.
Tube length = 12 ft
Number of tubes wide = 12
Parallel pass = 4
Number deep = 28

Determine the economizer performance. The tubes are in a staggered arrangement, and solid fins are used.

Solution

Key in the program, a listing of which appears in Fig. 5.3. In the RUN mode, the screen asks for all data in the same order given here; they are fed in. Then, using the logic discussed, the computer arrives at the solution. It is seen from Fig. 5.3 that the energy transferred is 11.2 MM Btu/hr, exit gas temperature is 273°F, exit water temperature is 458°F, gas pressure drop is 3.45 in. WC, and waterside pressure drop is 13.3 psi. The overall heat transfer coefficient is 8.04 Btu/ft^2 hr °F.

Table 5.3 gives the nomenclature, using which other data may be obtained.

Appendix

Correlations for Superheated Steam Properties

$$C_1 = \frac{80870}{T^2}$$

$$C_2 = 10^{C1} - \frac{2641.62}{T}$$

$$C_3 = 1.89 + C_2$$

$$C_4 = C_3 \frac{P^2}{T^2}$$

$$C_5 = \frac{372{,}420}{T^2} + 2$$

$$C_6 = C_5 C_2$$

$$C_7 = 1.89 + C_6$$

$$C_8 = 0.21878T - \frac{126{,}970}{T}$$

$$C_9 = 2\,C_8 C_7 - \frac{C_3}{T} \times 126{,}970$$

$$C_{10} = 82.546 - \frac{162{,}460}{T}$$

$$C_{11} = 2C_{10} C_7 - \frac{C_3}{T} \times 162{,}460$$

Using the variables C_1 through C_{11}, v, H, and S have been evaluated.

$$v = \left\{ \left[(C_8 C_4 C_3 + C_{10}) \frac{C_4}{P} + 1 \right] C_3 + 4.55504 \frac{T}{P} \right\} 0.016018$$

$$H = 775.596 + 0.63296T + 0.000162467T_2 + 47.3635$$
$$\log T + 0.043557 \{ C_7 P + 0.5C_4 [C_{11} + C_3 (C_{10} + C_9 C_4)] \}$$

$$S = \left[\frac{(C_8 C_3 - 2C_9)C_3 C_4}{2} - C_{11} \right] \frac{C_4}{2} + (C_3 - C_7)\, P \;\; /$$
$$T(-0.0241983) - 0.355579$$
$$- 11.4276/T + 0.00018052T - 0.253801 \log P$$
$$+ 0.809691 \log T$$

where

P = pressure (atm)
T = temperature (K)
v = specific volume (ft^3/lb)
H = enthalpy (Btu/lb)
S = entropy (Btu/lb °F)

Index